# LIFE ON A BUDGET

Written by

Mookey Mae Phillips

authorHOUSE®

*AuthorHouse™*
*1663 Liberty Drive*
*Bloomington, IN 47403*
*www.authorhouse.com*
*Phone: 1-800-839-8640*



*First published by AuthorHouse 4/19/2010*

*ISBN: 978-1-4520-1339-8 (e)*
*ISBN: 978-1-4520-1337-4 (sc)*
*ISBN: 978-1-4520-1338-1 (hc)*

*Library of Congress Control Number: 2010905369*

*Printed in the United States of America*
*Bloomington, Indiana*

*This book is printed on acid-free paper.*

This book is dedicated to my daughter: Christina

I have great hopes that she may someday read the contents of the following passages to give her the "desire" and "motivation" to want to enhance her life. She deserves the opportunity to embrace life and to live it to the fullest without having to deal with the constraints of the "American Dream" and the imperfect financial world.

## Contents


| | | |
|---|---|---|
| 1. | "VACATION -Yes or No?" | 1 |
| 2. | We all have to eat | 9 |
| 3. | Couch potato – time to relax | 17 |
| 4. | Fly like an eagle…"Where are we vacationing?" | 23 |
| 5. | Bah Hum Bug…holiday shopping | 29 |
| 6. | Just say "Charge it"…I don't think so | 33 |
| 7. | I don't want to cook tonight | 39 |
| 8. | Is it time to move? | 43 |
| 9. | Talk is cheap, but is the phone? | 47 |
| 10. | Rent vs. Buy | 49 |
| 11. | Chop chop…bye bye mortgage? | 55 |
| 12. | Teaching your kids about money and how it really works | 59 |
| 13. | Is it time to fly the nest? | 63 |
| 14. | Is there a way to teach your kids to save for the future? | 69 |
| 15. | Let's play dress up | 73 |
| 16. | Taxes…Do you know what you really should know? | 77 |
| 17. | Scamming your life away…Watch out for the bad guy! | 85 |
| 18. | Is your credit important to you? | 87 |
| 19. | Turning the nest into a castle | 89 |
| 20. | So, you want to tie the knot! | 91 |
| 21. | Miscellaneous things to keep in mind | 93 |

# PROLOGUE

Life is about making money, knowing how to save it and learning the suitable method in spending it so we can live an enjoyable fulfilling life. Our lives revolve around money whether we have it or we don't. Everyone's lives depend on having enough to survive and the "desire" to want more to be able to do the things that are important to us. The *want more* cannot come without "motivation". To have "motivation", you must have "desire". "Motivation" is not waiting for it to just drop into your lap such as winning the lottery or receiving a healthy inheritance. That does not happen very often to the average person. Most average people living in this country or even other countries for that matter must work to earn the money that we need to exist in what we call a normal world. We have the "motivation" to work so that we can live, but where did we bury our "desire" to envelop financial stability so we can live our lives to its fullest and enjoy every moment of the time that we have here on earth?

For many years, I have talked to friends, family and other people that I have run into about the basics of just living life at a simpler pace than trying to keep up with the Joneses. The feedback that I have received has been filled with both pros and cons. I have heard, "Oh, that is such a great idea!" to "That would never work for me". When I have spoken to them at a later time, I will ask them if they have taken what I have said into consideration. I usually get "I forgot about it because I figured it was too much work". Right there is a prime example of why many people fail. If you are afraid to succeed, you will never take a step forward to make a change for the better. Some will take a step forward, but when they realize that they don't want to change or that it is too difficult to change, they stop before they get far enough to recognize that there is hope. And that hope equals a better financial state.

Change is difficult for many people; what makes change happen is desire. Changing your life comes in a variety of different scenarios. Whether it is the desire to quit smoking, to lose weight, to quit drinking or to change your life financially, it means that there will be hurdles and none of these can be accomplished overnight. My mother always told me that Rome wasn't built in one day. Neither can major changes in your life be altered into something good in such a short time.

Change should be "motivated" by "desire" when you are dependent on living from payday to payday. Or you are interested in making a favorable financial change. There is always hope; make an attempt, and keep trying. All that is needed are attainable goals. I cannot make it happen for you; nor can you make it happen unless you truly want it to. I am hoping that by reading this book, you will gain some clear ideas on how to make an incredible life altering change for the better.

Many of the ideas in this book can be used by anyone who has a desire to change. It does not matter what income bracket you may be in or what you may be accustomed to spending. There are two words that you need to remember and they are "desire" and "motivation". There will be many different ideas and possible circumstances that are portrayed in the following chapters in which some may relate to your life and others may not. While some of the information may be useful to you, some of it you may want to share with someone you know.

There are no guarantees that the following information will work for everyone who reads this book. But, for those who do read this book, you may gain enough insight into how it really is not as difficult as one may think to change your life for the better. You need to revise your life to be able to make changes and have the "motivation" it takes to follow through until you reach the goals you set for yourself. Remember, that your goals need to be realistic and not so farfetched that you will never be able to attain them. Start simple and then work yourself up to larger goals later. Think of a baby who has just learned to take their first steps. Their goal is to get to the coffee table and get that something they shouldn't have. Their goal was not to run the 500k. So, just like growing up, it takes time to learn new things and for some of us it may take a little bit longer than others to digest the information in the following pages. I am hopeful that some of the chapters in this book will alter your life for the better. I am sure that not every subject that I cover may influence you into making a minor life style change that could be a long-term life event. Get the idea! So let's read on to see if any of this information can prove helpful in enhancing your financial situation.

# Chapter 1

## "VACATION -Yes or No?"

*Vacation is defined as "a period spent away from home or business in travel or recreation* ". Everyone likes "time off" from life to relax, play or travel. There are many people that I know personally who do not take time off because they can't afford to and if they do take "time off", they stay home because they don't have the money to travel. Do you have a dream that takes you traveling somewhere that you truly believe you will never see before you die?

Guess what! I have had that dream so many times in my life and my desire for going was so strong that it bummed me out because I thought I would never get there. My longest dream was my desire to visit Germany. Since taking German through middle and high school and having an ancestry mostly of German, I wanted to see what it was all about. In my early forties, my husband, Michael and I backpacked Europe (including Germany since he was born there) for eight days on a budget that was hard to believe.

We had talked about our European vacation for a long time and had a lot of trials figuring out that it was something we really could not afford. At the time, we had three children still living at home and their daily financial needs seemed to supersede the ability for us to go anywhere that extravagant. It was more like which state park shall we camp at during the coming summer.

It took a lot of "motivation" once we recognized that our "desire" was too strong to let go. We devised a grand plan that we knew was fool proof as long as we followed through. It took us a little over a year to save the money that did not affect our budget heavily. We cut back enough that it did not drastically disrupt our lives or even enough that our children really noticed. The following is the plan we implemented and followed through on. We did not deviate from it even once. The startup cost was inexpensive while the investment was worth every nickel saved.

The first thing we did was to go shopping. Remember, sometimes you have to spend money to make money. We started at a discount store and bought a ten- gallon fish aquarium. (No, we did not raise or breed fish… lol) Next, we headed to the hardware store and bought a small tube of bathtub caulk and a piece of Plexiglas (*defined as "polymethyl methacrylate - a transparent plastic used as a substitute for glass")* cut to fit the top opening of the aquarium. We took our supplies home and got to work putting our idea into action. The first thing we did was to drill a hole in the center of the Plexiglas that was big enough for a nickel to pass through. We opened the caulk and spread it around the top opening of the aquarium. It was then time to set the Plexiglas in place so the caulk would hold it secure. Perfect. We just needed to let it dry so no one could get the Plexiglas off. We were so proud of ourselves because we finished the first step. This was the easiest step of all. We set the aquarium in the kitchen on the counter so that we would see it many times a day. By seeing it often, it gave us the "motivation" we needed to get started and stay focused on our goal.

Our next step was close to the hardest of our goal. Every two weeks on payday, my husband and I went to the bank and bought forty dollars in nickels. We each took twenty dollars from our paycheck to put towards our savings. We then brought our nickels home and opened them and slid them through our "nickel" hole in the aquarium. If you are thinking, "Why nickels?", then think about it for a second. Are you willing to rip off the top of the aquarium to go nickel diving? Are you going to take a handful or two of nickels to the store for milk or to get gas? What about the kids? Ours would have been embarrassed running around with a lot of nickels. If you said "No", then we are on the same page. If you said "yes", I hope you will read this book from cover to cover so you will also learn about self-control.

The first couple of months proved to be disappointing because the nickels didn't look like much sitting in the bottom taunting us. This is when we began adding change that we acquired (after it was turned into nickels). There were several times that we questioned our motives and each of us was strong for the other. Any handyman chores that my husband received compensation for and anything I earned from taking surveys and taste tests was turned into nickels and put into our magical aquarium.

After this, our pyramid of nickels really began to grow and soon we were even unable to slide it from place to place on the counter. At eleven months the aquarium was almost full! We both agreed that it was time to really start planning our trip. Having spent close to a year of saving and agonizing over whether we made the right decision on a trip out of the country, we decided to plan our trip very cautiously with finances in mind. We applied for our passports first to give us the reassurance that this was real. I spent many anguished hours on the Internet researching airline fares. Did we want to fly into Amsterdam, Madrid, Dusseldorf, Brussels or Paris? I checked every possible airline site for every possible city in and near Germany. During all of this searching it occurred to me that we would need to get around once we arrived. I found that rental cars were too expensive and driving in a foreign country did not appeal to either of us. I did my homework and discovered that Europe relies on mass transit for transportation. It was time to research the Eurail pass (Eurail.com). There was a new dilemma to face, which meant we needed to decide how many countries we wanted to visit. There was Benelux, which included Netherlands, Luxembourg and Belgium. Then you could add nearby countries onto that or choose others that were more fitting. Before booking our train passes, I found a great flight from Minneapolis to Amsterdam which came in close to one hundred dollars cheaper than the other cities. So, I just saved two hundred dollars for our two tickets on Northwest Airlines. I then went to Eurail.com and booked our train passes that would cover us in Benelux, France and Germany for eight days for less than the cost of a rental car in one country for eight days.

We were getting close to turning our nickels back into paper money. We looked at the aquarium and identified with the fact that it was not going anywhere. This thing was going to break into a million pieces if we even tried to shift it even just a little bit. We started collecting small boxes to put the nickels into for easy transport to the bank. Another thought crossed my mind. Once we were at the bank with our stash of change, would there be enough room in the coin counters to take our entire load? I took a quick trip to the bank that we did business with and conferred with the bank manager. She would need to make a call to have the security team on location to empty the machine as needed. We set an appointment for the big event and waited in earnest to see how much money we had saved.

We were very excited as our car was loaded with all of our boxes and several grocery bags jingling with coins. The bank was not far and we arrived safely without spilling any of our filled boxes. Two hours and fifteen hundred dollars later, we hugged each other joyously as we embraced the reality of what we just did. We were going to Europe and get to see Germany too! Oops! Our trip is totally paid for but we don't have much left for eating, sleeping, attractions or souvenirs.

We sat down later that night and came to the conclusion that we did not look at the big picture even though we did think out of the box. We decided to pull a double whammy! We had already filed our income taxes and they should be arriving at least a month before our big trip. We decided that we would take our tax return and pay off our credit cards as we usually do each spring (This we will cover in another chapter) and then take five hundred dollars of that money with us and take one lower balance credit card along for backup.

Our trip was exceptional. We chose to take only backpacks (our kids made us the laughing stock because they explained that we were too old to travel with so little and that parent's could not backpack except for camping.) We did wear money belts under our clothes for the added security and knowledge that our passports and cash were safe. We wore fanny packs around our waists to carry the camera, extra batteries, gum, and a pen and of course, my cigarettes.

We survived the flight overseas and landed in Amsterdam and took the train into the city. We walked and did some sightseeing and planned what we wanted to do. We chose overnight train travel to save on hotel accommodations and it gave us the experience of many insightful situations while visiting abroad. Since we were very busy while visiting, we usually had bread (after three days we finally found butter) for breakfast, many times we skipped lunch and just had a small snack and then found some place decently priced for dinner.

We chose not to buy souvenirs for friends, but picked up a small item for each of our kids. We decided that this was our vacation (the first real one without the kids) and why should we spend all of our money on other people. It may sound selfish but it was our trip and not theirs. When other people laboriously saved their money for as long as we did, then they too could book a vacation and bring their own souvenirs home with them.

We tallied our total cost after returning back to the real world and found that our entire trip cost us less than two thousand dollars for both of us. This included airfare, train, hotel accommodations, food, entertainment and small cash advance from our credit card. We also found one hundred Euro while in Amsterdam and turned this back into the American dollar after arriving back in the states. I then applied this to the credit card that I had taken a cash advance from.

There are many different approaches to finding the best flight arrangements that fit into your needs. By searching via the Internet, there are countless websites that can lead you to the most suitable online company. Many people like to search by a travel search engine or research through a certain company. Personally, I prefer both. I always start my search for flight arrangements through a travel search engine such as Expedia.com, Travelocity.com or Travelzoo.com. (These are only a few of the numerous engines available.) I like this type of search because it gives me the opportunity to compare flights from one airline to another. I use several search engines to get a comparable idea. (The results from one search engine to another are usually very similar.) After I have found the "best" airfare, I always visit that airline's website to see if they are offering something "better" than what I found elsewhere. It can usually save you a booking fee by booking directly with the airline at their personal website.

Living in the Twin Cities in Minnesota can cause some grief when booking airfare as well. The flights seem to be more expensive to fly in and out of than some of your larger cities. If your travel dates are flexible as well as departure and arrival times, you may have a better chance for lower airfare by changing your plans ahead or back by one or two days. With this scenario of flexibility, Priceline.com offers a name your own price for airfare. If they accept the price you have chosen, you will fly by their itinerary. Your dates will not change, but you will not have a say in what time you will depart or arrive or even what airline you will fly with. I have found that I can save money on airfare by being a little creative. Several years ago, I needed to fly from Minnesota into New York City and was unable to find an inexpensive last minute one-way flight. (I needed to purchase tickets three days before my departure.) I checked and checked and was severely disappointed to find that my airfare would cost me close to eight hundred dollars. I began to panic and realized that if I was going to make it to New York, I was going to have to figure something out and do it fast! I brainstormed and had a vision hit me. I realized that flying by airplane was really no different than taking a bus. Busses make connections and that is what gave me the idea. I decided to check into flying one airline into Chicago (Chicago is a major hub and many times you can find better deals flying in and out of either Chicago O'Hare or Chicago Midway.) and flying a different airline for the second leg of my journey. I was pleasantly surprised. I found my flight from Minneapolis to Chicago Midway for forty-nine dollars and my flight from Chicago to LaGuardia was just under a hundred and thirty dollars. By using a little common sense, I saved myself over six hundred dollars!

Another way to get discounted travel or as I say "free travel" is to be aware of your situation when you are ready to travel. There are so many times that airlines over book flights and not everyone who has purchased a ticket has an opportunity to fly. Other times, people fly standby because an issue has arisen in their lives. The airline usually will offer a free round trip ticket within the continental U.S. if you volunteer to relinquish your seat for someone else. This means that you will fly "standby" without a guarantee to get on the next flight. Then once again, you will receive another voucher for a free ticket if you are not booked on that flight . There is a risk to this if you have set plans for destination pickup or a scheduled event. Do not try this if you cannot be in a position to arrive at your destination any later than originally planned. This can work in a situation when you are flexible, but may not be offered when you wish it would be.

Recently while listening to the Wall Street Journal news on talk radio, there was a short quip about traveling by cruise line. During the years of 2009 and 2010, the cruise line industry will be adding seven percent more ships to their fleets. The conversation also reverted to the fact that vacation cruises will increase in the future due to the low cost of booking. Depending on which city you fly from to the port city, you may be fortunate enough to get a great deal on airfare. Many of the cruise lines offer specials for ninety-nine dollars for a four night cruise. Most cruises are all inclusive, but where they make up their money is in the extras that they try to get you to purchase. Some of these extras may include spas and elegant dining. If you are on a serious budget, you need to remember to stick to your budget and just say "no" to the extras. You will still have a great time without adding the extra expense of a massage or elegant dinner.

Vacationing is usually fun, but if you don't stick to your guns it can turn into a disaster during or much later after your vacation has ended. By not being prepared and having made the best decisions, before and during your trip, it can cause side effects afterward that are both mentally and financially draining. Be very cautious of all of the details before booking and confirming any reservations. If you are not Internet savvy, ask a friend or family member to walk you through the process. Be very certain of dates of travel before booking as many airlines charge for cancellations or changes and this can seriously affect the deal that you may have just found. Try to fly on off days such as Tuesdays or Thursdays because you can get a better fare for flying during low peak periods.

If you are not in a position to afford going away on vacation because you do not have the funds to do so, a weekend trip to a local campground could be an alternative. (Keep in mind that if you do not already have the camping gear and necessities needed for a weekend getaway, you may need to choose another alternative so you can keep your budget in line.) A day trip to a nearby park or lake can be an inexpensive option when you do have money that you are not willing to spend. Find a location within an hour drive and pack a picnic lunch to walk along with you. (A backpack is great for this adventure as you can take your lunch with you and not have to carry cumbersome heavy coolers. Remember to pack bottled water for each person hiking. If you have several backpacks and have children that you will be taking with you, set up a light backpack for each child to carry with them. It will make them feel important that they can carry their own and be a part of this event.

Remember, that it is not worth ruining your budget for a vacation or trip that you cannot afford. By waiting and putting money aside for a vacation will be more of a fulfilling time that by putting yourself into debt or further into debt. Also, by waiting, you can plan all of the extra details that you might have missed if you could not afford them before. There is a time for everything and most times waiting will pay off.

# Chapter 2

## We all have to eat

Most people despise grocery shopping because it is time consuming. I am not one of them. Overall, I actually enjoy grocery shopping more than any other type of shopping. It is easy to make a game out of grocery shopping to see how much I can buy for just a little bit of money. I do not shop at a conventional grocery store such as Safeway, Cub Foods or Krogers. I have chosen to do ninety five percent of my household grocery shopping at a local discount warehouse type grocer.

I have found it very frustrating, time consuming and not cost efficient to shop at several different retailers for items that I can purchase at one location. If I have to go out of my way to save fifty cents on a product and that is all I need to buy, I will not divert my route, as I would probably have lost the savings in the cost of fuel. If I have to drive near the store for some other reason, I will wait until that opportunity arises and then stop and buy the items that I need at that particular store.

When planning a task such as grocery shopping, there are two things you should do before arriving at the grocery store. It is very important to make a grocery list of the items that you will need on this visit. You will find that you will spend more money if you do not have a list and you will purchase items that you do not necessarily need. The other thing to keep in mind and is imperative to do before shopping. Do not go grocery shopping if you are hungry. It is essential to eat prior to your visit. Have you ever noticed in the past that when you have shopped when you are hungry, you have a tendency to purchase more items and especially items that you would not normally purchase? I have found that even eating a slice or two of toast will take the edge off of my hunger and make it much easier to stick to the budget. Don't forget your calculator so you can financially stick to your budget and not over-spend. I have often put back items of less importance so I am able to purchase the items that I really need. I have a five-dollar cushion to cover tax and anything I inadvertently added incorrectly.

There are many discount grocers in most major American cities nationwide. I do not include shopping at places like Sam's Club, Costco and warehouse grocers that enforce a paid membership system to be able to do business with them. Unless, a person does most of their shopping at places like these; it does not always make sense financially to belong. I have never been one to shop in bulk, as I usually do not have the space for storage of large quantities. Many consumers do very well with this type of store and utilize their membership immeasurably, while others purchase a membership anywhere from twenty-five dollars to seventy-five dollars per year and rarely use it. I personally do not have any type of membership shopping plans as I rarely shop in bulk nor do I desire the additional fees that come with this format of shopping.

My number one choice of grocery retailer is Aldi Foods. There are no membership fees, but the shopper does need a quarter to use one of their grocery carts. When the cart is returned, the quarter is returned to you. This is a practical means of eliminating overhead for the company so they are able to keep the cost of their products at a lower rate, which in turn hands the saving over to the consumer. Aldi Foods does not bag your groceries for you either. Nor, do they provide free bags for the consumer use. Bags are available to purchase for a very low fee per bag or the shopper can bring their own bags or remove empty boxes from the shelves for the removal of their groceries from the store. This also eliminates overhead and the savings again can be passed on to the consumer. I have my own bags that I take with me on each shopping trip so I do not have to purchase any nor do I have to break down any boxes that I need to recycle after I unpack my groceries at home.

I have set a strict budget for grocery shopping at one hundred dollars every two weeks. I am able to purchase ninety-nine percent of the items that I need on each of these trips. On occasion, I may need to stop and buy an additional gallon of milk or a loaf of bread. This store is on my way home from work, so I am fortunate that I am still able to cash in on the lower savings by not having to drive out of my way.

I am cautious of what I feed my family and before I go grocery shopping, I have planned meals in mind. I do not shop when I am hungry, like most shoppers, I have a tendency on buying the things I am hungry for instead of keeping my budget in mind. Most of my shopping is done when I can do it by myself. I have found that my husband does not like to go grocery shopping and wants to get in and out of the store as quickly as possible. When he has gone with me, I have come to the conclusion that he will stay at home because I miss half of the items that I needed to pick up. I also prefer to leave my son at home because he will see something he wants and I have to try and eliminate something else to put his item into my one hundred dollar budget.

Most of the items that this retailer has for sale are equivalent to items that I can purchase at the top local grocery chains. The only difference in most cases is the brand name. There are a few items that I will not purchase such as catsup or bologna from my store of choice.

There is a wide selection of fruits and vegetables to choose from at a discount price and the fresh meat selection has enough options for me to choose from. One item that I do purchase in bulk is fresh ground beef. I prefer to buy a five pound package because I can usually save anywhere from twenty to thirty cents per pound. I like to divide the package into five one-pound freezer sandwich bags for easy convenient use later.

Peanut butter, jelly, pop tarts and spaghetti sauce are staples that I buy most often. I used to purchase off brand peanut butter at name brand grocers and have noticed that I can save almost seventy cents per jar where I shop now. When it comes to purchasing strawberry jam from local grocers, it is aggravating to see that the cost is nearly three dollars for the same thing I can purchase at my grocer for less than half of that cost. Name brand pop tarts taste very similar if not the same as off brand. Since my husband and son like them for a snack or for a quick breakfast, I regularly buy them at a savings of over one dollar per box. I have had a lot of people tell me that they will not buy off brand spaghetti sauce because they don't like it. When I cook, I usually add spices to my jarred spaghetti sauce regardless of whether it is name brand or off brand. I get the same desired results but at a much lesser cost than many others pay.

This type of store offers all of the essentials that I need including health and beauty aids, pet supplies, dry goods, baked goods, canned and boxed goods to frozen foods. By having a *great* budget, I have also cut back on buying junk foods that my family does not need or miss, which in turn provides me with the opportunity to purchase more food and prepare better meals. With healthier and wholesome meals and strategic planning I am able to keep my family happier at the dinner table. In addition, this aids in being able to set aside *extra* money for that vacation we want to take, new shoes for one of the kids, or rainy day money.

The varieties of different foods are endless as long as you, the shopper, have an open mind for changing your habits and routine. I can usually save between forty to fifty dollars per shopping trip as long as I keep my "desire" to save and my "motivation" for following through with my plan.

Be cautious of delivery services for grocery shopping. Many cities throughout the country offer Internet ordering for grocery shopping. There is usually a delivery fee depending on the amount of your grocery order and many of the grocery items are more expensive than at a local grocer. If you need to cut back, this option is not a cost saving one.

Coupons can assist in saving some money when shopping. Most major newspapers will distribute coupons each Sunday and depending on your product usage, they may be useful in helping you lower your spending for your shopping visit. Check with your local supermarket to see what their coupon policy may be and to see if they have a "double" coupon day. This may give you additional savings on certain "cents off" coupons. Years ago, I actually changed my product usage so I was able to utilize double coupons. It was not unusual for me to save fifty to seventy-five dollars per shopping trip. When the grocery store changed to "triple" coupons, I was able to save close to one hundred and fifty dollars per visit. Look to see if the product is on sale, which will save you even more money when using coupons.

Be a smart shopper while keeping your family's health and budget in mind. It is best to shop when you are not hungry and when you have a list in tow. It is easier to stay on the beaten path if you already have your plans in order and have the proper tools. Remember to stay firm if you really want to reach your goals.

There is another alternative in saving additional money. When grocery shopping, be committed to your plan of saving money. You are at the grocery store and suddenly you reach the produce department; what do you think you do? I am guilty as well and used to do it all of the time.

Because of convenience and the need for ready meals, you reach for the bag of lettuce mix that is completely prepared and ready for immediate use. The cost of this product overall is much greater than as if you were to purchase lettuce, cabbage and carrots separately and make your salad from scratch at home. This applies to other products as well. Baby carrots can be purchased peeled and ready to eat, but that cost will also supersede purchasing a bag of carrots and peeling them yourself. There are many tricks to learn when it comes to grocery shopping and cooking at home. During the right season, depending on your growing conditions, it is not difficult to grow herbs and spices that can later be dried in the oven or used fresh. Even a small vegetable garden can save a bundle on your trips to the produce department. This does not mean having to invest in a new hobby. Just grow what you think you may consume and then see what you can save in the long run. Let's say that you end up planting too much and your harvest is greater than what your needs are. There are a few options to counteract the outcome so as not to waste the remaining vegetables. If you have time to blanch your vegetables, they can be frozen for use at a later time or canning your extras could be another option. If neither of these choices will work for you, offer some of your extras to friends, family or neighbors or integrate them into upcoming meals so there is no waste. Boneless chicken breasts are often much more expensive than what we would like to pay, but we do so because of convenience or not having the knowledge to de-bone them ourselves. There is so much information available on the Internet or at the library to teach you how to accomplish this. After you have practiced several times, it will seem like you have always done so. There are so many more items that you can prepare yourself rather than by ready-made. Just use a little common sense and you will soon start saving more than just cents!

Another tip at saving a little bit of money is watching how you purchase the items that you need. For many of us average consumers, buying in bulk may work great especially when you have enough family members that you are buying for. If this doesn't work for you, you may ask yourself how this might benefit you particularly when you can't see how it would. You may already believe that you are already doing the best for your budget when you shop and you may already be doing what I am about to suggest. First, you will need to review your monthly needs when you shop. Secondly, you need to decide if you are making the most of your budget and lastly, you need to change the way you think about shopping and planning if you can save yourself even a little bit of money if you implement a different strategy.

Even if you are only shopping for two or three and you are buying size appropriate packages, you may have an opportunity to enhance your budget by executing a different tactic. On a usual grocery-shopping trip you may pick up size appropriate packages of beef, pork or chicken to feed to your family that particular week. What if you were to think out of the box and plan for a longer period of time? The first thing you will need to do is to invest in either some zip type sandwich or quart size freezer bags. You can substitute freezer paper but this method is not as cost effective and may defeat the purpose of saving money in the long run. A true dollar store that carries these bags will only cost you one dollar per box and cut your spending. Next, instead of purchasing size appropriate packages that accommodate your lifestyle and needs look at purchasing family packs of meat. Family packs offer "cents off" incentives per pound. An example of this would be to purchase a ten-pound package of ground beef that is offered at ten cents off per pound. You will have just saved one dollar or more just on that package of meat. After arriving home to put your newly purchased groceries away, divide the package of meat into smaller portions into the bags you purchased from the dollar store. Your overall savings with the cost of the bags could still save you up to ninety cents or more for the package. The ground beef then can be placed into the freezer for use sometime in the near future. (Many cookbooks will give information on how long certain meats and vegetables can be frozen until it is no longer safe to do so.) By using this method with all of the meats you purchase, you will eventually be able to see how much money can be saved by shopping smart.

I have an example of how you can save by buying in bulk with health and beauty aids and household cleaning items. By purchasing bar soap in a larger package such as a six pack or ten pack, the savings can be quite significant than purchasing an individual bar or small pack. When visiting a true dollar store that only charges one dollar per item, it is typically very easy to find every day household cleaners, health and beauty items and kitchen items that you would spend so much more for had you shopped at a local discount store such as Kmart or Target. It may be more cost effective for a short time to purchase a different brand of shampoo or soap to help make your budget work. In the future, once you have caught up your finances or savings to where you need to be, changing back to your preferred brand may be an option.

When shopping at discount retailers such as Target or Kmart, check the clearance items to see if you can locate a great deal. On one of my recent trips to Target, I looked at the clearance health and beauty aids. I was thrilled to find a combo pack of four liquid hand soaps for $3.38. With a quick calculation, I realized that I would save close to seventy cents (including the tax savings) if I purchased the item right away instead of buying it at the local dollar store.

Remember a dollar saved is a dollar earned. Had you used that savings towards a purchase elsewhere, you would not be able to use it towards something you may need at a later time. Frugal is defined as: characterized by or reflecting economy in the use of resources. Being frugal is not a bad thing as long as you do not allow it to control and over take your life. When this happens, the compulsion to be frugal interferes with normal tasks and events in your life and may cause issues with family situations or possibly even depression. Always keep in mind to have a healthy relationship with your daily budget and you will eventually see the improvement of your financial position as long as you have "desire" and "motivation".

# CHAPTER 3

## Couch potato – time to relax

I don't know of anyone who doesn't like some form of entertainment, whether it is watching television, renting a movie, reading a book or even just surfing the Internet.

Let's start with movie watching because this can be a big expense to a budget that may be close to falling apart. Renting movies can be hazardous to the consumer's pocket if they are not cautious on what they spend on rentals or even buying new release movies. New release movies usually range from fourteen dollars and go up from there. If you buy a couple of them, you could be spending thirty dollars or more depending on the price tag. Or a night out at the movies with two paid tickets (if you leave the kids at home with a sitter), popcorn and beverages could end up costing thirty to forty dollars. Let's keep adding. Don't forget that you have to pay the sitter for watching the kids. There goes another ten to fifteen dollars. So, that means the total for a night at the movies could end up costing fifty dollars or more. Ouch! Yes, some movies should be seen on the big screen, but you don't need to destroy your budget by seeing them all. Wait for them to come out on video and it will save you a handful of money. If there are movies that you don't want your children to see, put them down for the night before starting the video or wait for them to have events with friends that will put them away from home. Most cities have discount theaters with seats costing two or three dollars per show. You may need to wait several months for them to play at this theater, but you will still be able to see that special movie on the big screen.

The American population is in a big hurry to do what we want and see things as fast as we can. Be different so that you can add another good change to your budget. I don't think that it is too difficult to wait for a movie to come out on DVD or to come to a discount theater. Waiting for the movie to come out on DVD can give you immeasurable savings if you utilize your "desire" and "motivation". It also teaches you and your children that it is okay to wait because this is only a movie and not a priority.

Remember that your entire family can see that movie you have been dying to see at the theater for a fraction of the cost by waiting for it to become available on DVD. The savings are immeasurable when you consider the out of pocket expense to purchase two or more theater tickets, the cost of something to drink and something to munch on such as popcorn or candy or both. By renting the DVD, you can save the admission price of theater tickets and provide your own snacks and beverages for a fraction of the cost.

Renting movies from online sources can save you money if you rent as many movies as you are able to on the plan. If you do not, you will not save money. A lot of the online rental movie plans that come through the mail, do not always have new released movies in stock and the renter needs to wait for that movie to become available. Renting at movie stores such as Blockbuster, Hollywood Video or Mr. Movies can be expensive if you rent movies a lot. The last I checked, a rented movie from this type of store ran at least five dollars. I have an alternate option for renting movies at a much greater savings and can get you free movies as well.

Most of us have seen the "Redbox" located inside McDonalds and some grocery chains. Their system is quite simple. There are easy to follow directions on how to rent, return, purchase used and to rent with a *promo code*. Their movies rent for one dollar and needs to be returned by the following evening at the posted time on the machine. (It is usually seven pm, but I have seen nine pm also.) You must return the movie by the posted time or you will be charged an additional day which the cost is one dollar. The only drawback is that a credit or debit card is needed to rent from the machine. There is a great opportunity here to save money with free rentals. There is a website called www.redbox.com where you can sign up for a weekly text message to get a free *promo code* for a rental to be used on a certain day. You will also receive a free *promo* code when you sign up with your email address. You can also go online and Google a search for free Redbox codes for the state you reside in. I have used DVDONME, MILLER, BREAKROOM, and DESSPAD. Many of the codes are state specific, so check your codes before you use them. Or you will find out when you do use them that they are not valid. You may use your *promo code* only one time per credit card. If you have three different credit/debit cards, you may rent three movies with the same code and get each movie free. Keep in mind that the movies need to be returned on time or a dollar late fee will be charged. Redbox movies can be rented from any Redbox location and returned to a different Redbox location anywhere.

Another great place for movies is your local library system. There are hundreds or more movies available in many different genres. By visiting your local library, you can see what movies are on hand to borrow. Check their catalog to see what movies may be out or at other locations. Most movies through the library can be reserved and the library will notify you when they are ready to be picked up. Most libraries allow you to check out the movie for close to a week period.

You may like to watch television but either don't have one, you get poor reception, or have only one television and two people in the household who want to watch two different programs. If you have Internet access and have at least broadband or cable, you may be in luck. This means you may not need to go out and spend hundreds of dollars on another television. Many of the television sites such as ABC have a link to be able to watch a primetime television show right on their website. You may need to download a movie player that is site specific, but it may be beneficial in the long run. The only disadvantage to this is that you may need to wait a day or two before the program is uploaded to their site. Due to my schedule last year, I was unable to watch Lost on primetime so I just waited two days and watched the program on (ABC.com). It was nice because I could start and stop the program at anytime and come back to it when I was ready. I was thrilled to have been able to keep up on one of my favorite shows and not have to worry about the possibility of not being available when it aired on Primetime or that I might miss recording the show for viewing at a later time.

Eleven years ago, my husband and I came across a great savings plan that saved us hundreds of dollars. We purchased a house and decided that when we moved in we would wait to call the cable company to have them install it. With involving our children in the move and preparing everything in our new home, we found that the children didn't complain much about not having cable. So, we went without. Occasionally, someone in the house would complain about how they wished they could have seen this program or that program, but overall it worked well for us. Over time, we found that our children became more involved with other things and didn't seem to miss it as much. Twice, we have tried the satellite deals and both times we got stuck into a year contract and also that we didn't seem to watch any of the channels often enough to make it worth having. If the need does arise to watch something that we normally might not be able to watch, we will turn to the Internet to see if we are able to stream it online.

The cost of buying books can add up to a lot of money that could be saved for other necessary things. Several years ago while dusting my bookshelves, I realized that I owned a lot of books that I had read, some that I would probably never read and still owned with a layer of dust on them. This is when I realized how foolishly I had spent my money. I took all of the books that I knew for a fact I would never read again and drove them over to the local library and donated them. Two things happened. The first was that I eliminated the books I didn't need while freeing up space in my home and the second thing was that I received a donation to write off for my taxes. I killed two birds with one stone. From that day on, I only purchased books that I would never want to part with. I borrowed books, CDs, DVD's and audio books from the library. The new release books took a while to get, but if I put myself on the waiting list right away. The wait was not long and was worth it when I realized how much money I was saving. If I were anxious to read a newly released novel, I would drag my family to the local book- store (I had no resistance) such as Borders or Barnes and Noble and find the book I was looking for. I did not buy the book, but instead took the book into the book- store café and sat down and read what I was able to read in the amount of time that I had. Sometimes, this book would show up just a few days later at the library and I could pick up where I left off. Other times, we would take another trip to the bookstore so I could read more of it.

Those of us who own a computer or who have access to a computer understand how useful they can be and realize how much information is in the Internet world. But the question here is do you utilize the Internet to save you money? Research can be unpleasant if it is something forced upon us. But when research can be used to expand our knowledge on product information and pricing, the results can be very valuable in helping us make better choices for our budget. Instead of driving to three or four different stores and wasting fuel to find out which retailer has a better deal, go online and check prices. Most major retailers and many small retailers have their own website and have a lot of information posted. So, not only do you save on fuel, but you find the best deal too. Don't forget to check out a retailer's website to see if they have a clearance section. You may find something different that you could settle for and spend less money. Another means to saving money with the Internet is to do some research on who provides the best Internet service and who can bundle two or more services (i.e.: local telephone service, Internet or satellite) to save on separate billing. This would be another way to save additional money for other expenses or something we "desire".

Please keep in mind though and look at the big picture. Do you really want to be a "couch potato" and go through life watching other people's lives real or fictitious and waste your own? Television was originally designed for passing on news and for entertainment, not for babysitting ourselves when we want to be lazy. This has caused many consumers to become procrastinators and put off doing something they need to do. When vacationing, unplug the television set in your room. You are there for exactly that...vacationing. You can watch your favorite show when you get home-it is also much cheaper than the cost of a hotel room for one night. Money is not always plentiful and is not easy to come by so use it to the best of your ability and treat it well.

The Internet is a good resource for keeping the whole family entertained. There are many television sites to watch previously aired programs to gaming sites such as slingo.com or pogo.com to video sites such as youtube.com.

The availability of these resources definitely outweighs the cost of regular subscriptions to spending money going out on the town with money that should not be spent. Look at the big picture and think outside of the box to maximize your savings that will assist you in making your budget work.

# Chapter 4

### Fly like an eagle…"Where are we vacationing?"

My husband and I have been questioned many times about how we can afford to take the vacations that we do. We are middle class and have three children but have learned over the years that we do not have to compete with Middle America. We do not need to have the newest or the best to feel as though we lead a good fulfilling life. I am not saying that I do not need or want new things. We do not need them to make us happy nor do we need the show for status quo. Many times a simpler life is easier and less overwhelming than trying to keep up with everyone we know or don't know for that matter.

Perseverance has taken a major role in our decision regarding travel. "Desire" and "Motivation" is always the key to anything we do or want to do especially if we are looking for favorable results. Most of us have more time than money, but time can also be sparse in our busy lives. Utilize your time wisely and do your research in depth to get the results you are looking for. In many cases, planning a vacation can be difficult alone without adding in the factor of money.

There are many websites available on the Internet to do enough research on where you may want to travel. When planning a vacation, I always look at alternative locations in the event that one of our choices doesn't play out financially. This meaning "We are not going there!"

I signed up for the E-club at travelzoo.com and receive weekly-generated vacation deals. This is where I found a trip that my son and I took to Japan. Several years ago, the email arrived in early January for travel to Osaka, Japan at the cost of one hundred ninety nine dollars each way, but I had to book by the end of that day. Travel had to be completed by

April first. I booked two tickets which cost me nine hundred seventy dollars round trip including taxes. I had a lot of research to do from this point on. I already had a passport from traveling to Europe, but I had to have my son apply for his. Unfortunately, we had to have it expedited. If I could have seen into the future and this possibility existed, I would have applied for it a long time ago. This would have saved me somewhere around fifty dollars that I could have used towards expenses. I went online to research accommodations, transportation and sightseeing. I researched mass transit since I knew that Japan had a great train system only to find that I needed to purchase our train passes here in the states as they were not available at discount once we were to arrive. I found a local travel agency that participated in selling the passes and purchased them for less than five hundred dollars for both. (Had we waited until we had arrived, the cost would have been astronomical for the amount of time we spent traveling from city to shoreline to city.) I researched hostels because I knew it would be much cheaper for us on our budget and it would also be more entertaining. Overall, for six days traveling by train throughout mainland Japan, staying in several hostels and hotels, choosing our meals and entertainment wisely, our trip cost us less than twenty-two hundred dollars. We enjoyed every minute of it and I am glad that we had the opportunity to travel there. My son learned a lot about architecture, food, art and culture. We saw a completely different world and we wished we could have brought home many of their concepts.

Now it is time for you to do your own research for your "desired" vacation. Get your "motivation" in gear and let's get started. First you and your family (or just yourself, if you are traveling alone) need to make a few decisions on where you would like to visit and have some alternatives in case your first or second choice is too pricey and does not fit into your budget. (By this time, you should have cashed in your nickels to see just how much spending money you have for this trip – hopefully, you did save nickels or had a savings plan in effect so as not to overcharge on your credit cards or use them at all.) You will want to do some research on sightseeing and things to do in the cities you have chosen to give you an idea if there is enough there to keep you occupied or entertained.

Now is the time to surf the Internet for airline fares. There are many search engines for searching for fares. You could check out sites such as expedia.com, travelocity.com or bookingbuddy.com. There are so many search engines that if you do not like my examples, you can Google a search to find a search engine that you may be happy with. After checking out the popular sites, I always check out individual airlines to see if they might be running a special which is not posted on one of the search engine sites. I have also learned that by booking directly with the airline carrier, you can avoid paying an online booking fee. Minneapolis is an expensive city to fly out of, which is where I live, so I will often check out airtran.com for travel in the east and southeast part of the country. They offer low priced ticketing with no Saturday night stay and no two to three week advance purchasing. (You can also use two different airlines to schedule your trip. It may be cheaper to fly to one destination, catch a different airline and fly the rest of the way') If your vacation time is flexible, it definitely pays to look at alternate flying days for your trip. Just by changing one or two days can save you up to a hundred dollars or more. Southwest Airlines recently began flying from certain cities in the Midwest and has been offering phenomenal fares to many different cities. They can be located on the Internet at southwest.com and they are not included as part of a search in the travel search engines. So, do your research well. Surfing and researching for several hours is priceless when you can save yourself or your family a big chunk of money on airfare, which could amount in the hundreds of dollars. And it gives you more opportunity on your vacation because you have loosened up some cash flow.

By now, your ticket is most likely purchased. It is now time to start thinking about whether you will need a rental car at your destination. And if you do, it is best to get back into your seat at your computer and settle in for another round of searching. Keep in mind that if you have a company discount on a rental car or even coupons, this is not always the best choice in making reservations. Many companies believe that if the consumer has a coupon, they will not check the competition for a better deal. Coupons are not gold; they are just another form of advertising to lure the consumer into using their business. This is where my research gets pretty hot and heavy. I check out the websites that offers my corporate rates, check the individual car companies and then go one step further. There is a website at priceline.com and they offer both a search for popular rental car companies and also give you the opportunity to "Name your own price". You will need to supply all of your information including a valid credit card that will be charged if your "price" is accepted. My preference for saving Priceline for last is because in my research I am able to see what the average rental car rates are. With this information, it gives me an idea on where to start my bidding. I always start low at less than ten dollars a day (I usually start at $6) and then work upwards from there. One weekend, my husband and I rented a car for three days from our local airport so we could drive to Chicago and not put the miles on our own vehicle. I started bidding at six dollars a day and finally had an accepted offer at eight dollars a day. When I picked up my rental, the car size that I had reserved was sold out. The rental car company upgraded me to a larger vehicle at no extra charge. What a savings. Granted, you still need to pay for taxes on top of the rental rate, but many times there are extremely good deals out there and you will pay less than the advertised rate.

Making sure that you have a place to stay is very important. I know that I do not want to search for a hotel as soon as I get to my destination and my preference is not staying near the airport. Okay, by now you are getting used to the idea that your computer chair is the only seating right now. On the Internet, there are several websites to visit to find hotel accommodations. My favorite is hotels.com because I seem to always find the best deals that meet my needs. Depending on how often you travel and use their site to book, they offer incentives for future discounts on your stay. I have used this site especially when I am unfamiliar with an area because it offers me more than what I would expect from this type of website. Each hotel that is available for your travel dates will appear with a price, stars (awarded on a one to five scale of good to best), complete hotel information, and best of all the distance from the main airport. Another site to consider is hotwire.com. Most hotel chains have their own website and if there is a particular chain that you like to stay at, check to see if they have a website. Compare their prices with what you find on the search engines. Remember that a hotel/motel is basically only to sleep. So, it doesn't need to be overly fancy. My thought is "If it is clean, inexpensive, offers free continental breakfast and is centrally located, I am happy." Free breakfast saves me money an time.

Packing for a vacation is very important. Airport security is still enforcing liquid limits for carry-on luggage. (Carry-on luggage is the only luggage I take so as not to incur additional charges for extra checked baggage and I am never in a position to have my luggage "lost". Laundry can always be done at your destination and this way you get to pack a little bit lighter. If something doesn't fit into my luggage, I re-evaluate the contents and eliminate what I really can go without.) I personally have stopped taking unnecessary liquids with me unless it is crucial. It is very easy to stop by a drug store in the vacationing city and pick up shampoo and the likes that you need. You have made it to your destination with saving on airfare, finding a lower priced rental car and an economical hotel. The last step is to have a good time and don't forget to keep your budget in line. Make sure that you have budgeted spending money into your vacation or weekend getaway so you will not be financially burdened when you return home. Thus meaning, you have travelled in vain.

Another option for saving money on airfare is to have a frequent flyer card. Most airlines offer them. Each time you fly with that airline, you receive a point for every mile flown. Many of the programs have incentive programs as well to increase your points toward airfare as well. Depending on your credit stability (meaning the ability to be approved for credit and the ability for you not to use your credit card to put you into debt), there are many promotions available from several credit card companies to earn free airfare by applying for certain cards. I have done this several times where I have applied for a new credit card in exchange for receiving a free domestic trip (25,000 miles). I use the card once (my offers have instructed that I need to use the card once before the miles are available for my use) and purchase a bottle of soda or a pack of gum and pay off the balance and put the card away. I have one card that I have only used once in the past two years and sent my son on a trip to see his cousins. My cost was ten dollars for the process of setting up the trip through my frequent flyer card. If you are not good with credit cards, I would not use this as an option as it could wreak more havoc for you and cause you to become more in debt. If you have a good sense and have the willpower to not utilize your credit card to cause debt, this could be a very viable option. You need to weigh the pros and cons for what is best for you. If you are unsure, please read the chapter in this book regarding charge cards to help you understand what the repercussions could be for you.

# CHAPTER 5

## Bah Hum Bug…holiday shopping

Earlier in this book, you learned that overall I hate shopping. I do shop and sometimes I really do enjoy it but I do not like busy stores, especially when people start the pushing and practically run you over with their cart because they have to be one step ahead of you to get that last discounted item on the shelf.

Holiday shopping is a good time to look for deals. Black Friday, which is the day after Thanksgiving, is actually the opener of the holiday shopping season. Why not research some of those deals before they appear in the paper the day before the sale begins. You will need to go on the Internet to do a little bit of surfing. Yes, again! You will need to "Google" search *Black Fridays Ads.* There are several different sites to choose from and some of them offer email sign up so you can receive the latest BF ads right in your inbox. Most of the stores open early while one or two of them may even be open on Thanksgiving Day which gives another opportunity for latching onto a great deal.

Where has your budget gone? Do you really need all of those things on your list? How long are you really going to stand in line to wait in hopes of getting that something you think you can't live without, but will discard in three months? One year, my husband and I were in Denver over Black Friday and waited in line for almost three hours. It was cold and I was ready to leave. We made it into the store and the salesperson quickly explained that the DVD players were already sold out (How does that exactly happen when they were supposed to be prepared and none had been sold yet?) Thankfully, he went on to explain that they would be substituting a different model that would normally cost more. The DVD player did have a rebate that was supposed to be sent in, but it would not apply to this purchase. We purchased two of the DVD players and a few other items and went on our way. We talked about how little we had spent and were happy with our purchases. We did not realize what had happened until we returned to our hotel. After looking at our receipt, I understood exactly what had happened. The sales person had to key in the purchase price, but while doing so entered the actual sales price of the item and then discounted it. He was supposed to have entered the original price and then discount it from there. The sales price of the DVD players was a good deal, but after the salesperson made the error, it was an incredible deal that I am sure only a few others saw before the error was corrected.

My husband and I rarely hit the Black Friday sales due to the overwhelming quantity of people who somehow show up long before we get there. Our favorite bargain shopping is the week or two after the Christmas holiday. Many stores discount their decoration inventory up to fifty percent off two weeks before the holiday. Many times, we have struck a deal for the "emergency" last minute deal. After the holidays when I visit a certain retailer, I will watch closely for their "clearance" sales. When the sales reach between fifty and seventy-five percent off, I get into shopping mode. This is a great time to start holiday shopping for the next year. I always keep in mind that there may be expiration dates for food products and consider how well they will store until the following year. Gift baskets always come in handy and I look for the type where they are not holiday related in the event that I need a surprise gift during the year. I am cautious of purchasing clothing, because trends can change quickly.

Many of the clothing stores will have a blowout clearance sale after the holidays as well. If and only if I am in need of clothing for the family will I stop by the sale. Be careful, many retailers will give an extra ten to fifteen percent off if you use or apply for their credit card. This is something very serious to consider. If you do follow this path, will you pay the balance off when you get your statement or will you only make a minimum payment due? By only making a minimum payment due, interest charges accrue until the balance is paid. You could possibly pay more for your purchases and not even see that discount clearance savings if you do not spend or charge wisely. Shop wisely with your budget in mind and remember that if you really don't need it; don't buy it!

With the economic challenges that we currently face, it is very important to keep your budget in mind. It is very important that you consider your "*desire*" and "m*otivation*", *before* committing yourself to unnecessary purchases. It is very important to learn to break the habit of impulse buying. When shopping, ask yourself if you really need the item(s) (there is a difference between need and want – if you do not know the difference, learn it and always separate the two whenever there is money to be spent.) and if you don't, put it back and walk away. It is all "*desire*" and "*motivation*" that will change your financial life for the better. A learning experience for my son when he was younger has turned him into a cautious spender today. Years ago, it seemed that my husband and I would drag the kids with us for another unnecessary shopping excursion to our local Target store. My son would frequently ask me if he could borrow money until he got home so he could buy some meaningless toy. Finally on one visit, I told him that I would not lend him the money and that next time we came to the store he could purchase the item if he had his money. The funniest but greatest thing happened the next time we visited the store. I asked him if he had his money and he told me that he did. I reminded him that the last time he saw the toy that he had wanted and did he want to buy it today. I was so thrilled when he said, "No, mommy, I don't want that toy anymore." I played this game with him many times until he no longer asked to borrow money. Today, he is a high school senior with a driver's license and a car. He is very cautious with his money, takes into consideration whether he really needs that special something and also considers the cost of fuel for his car before he goes anywhere that is not necessary. Did I raise a miser? No, he still spends money but ninety five percent of the time puts a lot of thought into any purchase that he makes. I am hoping that this event will enhance your child's life. You just need to implement it for it to work and you need to stay firm by staying on the path and not giving into tantrums or arguments.

A visual lesson can be well learned as long as you pay attention to all of the signs. With our struggling economy, many of our retailers are going out of business and are advertising incredible deals. Those incredible deals can be very misleading. In Minneapolis, I have visited several of these GOBS (going out of business sales) only to find disappointment while I browse the shops. Some of the GOB retailers offer great discounts staring at twenty percent off and progress up to seventy-five and occasionally ninety percent off. Be wary by knowing your prices before you purchase. Two stores that I have recently visited are in the final stages of closing their business and I was appalled by what I saw when I was there. Since I have done a lot of browsing at both stores, I was very familiar with their prices on many of their items. At one place, I saw a Crockpot that would make a great replacement for my ten year old one that was on its last leg. I looked at the price tag and could not believe that the store had actually marked up the price! So, instead of getting a great deal, I would be paying close to the original retail price after the advertised forty percent off. That is not a deal. I believe that it was the retailer's way to try and make up sales lost on other items. When I visited a clothier that was closing its doors for good, I found the same thing had happened. I am not saying that every single item in the store was marked up, but what I am saying is to be aware of current retail pricing at other retailers to ensure that you will get the best deal. I did buy a Crockpot a week later for twenty dollars less than at the GOB store and do you want to know something? It is still in the box, my old one is still in the cupboard and I have had the opportunity to use the new one but have not. So, I guess I really didn't need that new Crockpot, huh? Even I still make mistakes in vain.

# Chapter 6

### Just say "Charge it"...I don't think so

Many Americans owe much more on their credit cards than they should or ever expected to owe. It is okay to have a credit card as long as you are responsible. Sometimes it is necessary to have a credit card for making airline, hotel or rental car reservations. But, it is not necessary to have more credit than what is needed. If a credit card company raises your limit without your request, more often than not it is best to contact your lender and ask that the credit limit be return to the original credit line or less if you do not need that much credit. An average American family should not exceed more than five to ten thousand. We already know the difference between want and need as we discussed this in a different chapter. There is no want when it comes to credit cards because they can be very dangerous and hazardous to your life.

Here are some important tips that you will need to know or what you don't know can cost you. Unless you want to be paying off your debt for a very long time, pay more than the minimum payment. For example, if you made a minimum payment on a one thousand dollar credit card balance at fifteen percent APR, it would take eight years and ten months and cost $1729.19 to pay off the balance.

Be sure you know what you are paying for, so review your statement carefully. Your interest rate and fees can change after you get your credit card; read the fine print and do not use cash advances because you will pay a fee of three percent of the amount you borrow, plus interest on the balance. If you use convenience checks, the bank may charge you a fee for the use of them. To protect yourself against identity theft, write to your bank and ask them to stop sending you credit card convenience checks.

If you insist on having credit cards, only hold cards with a no annual fee. Most credit card companies today no longer have annual fees. There are many credit card companies that offer "rewards" cards without fees. Beware of "default interest" because your interest rate can increase if you miss a payment, go over the credit limit or bounce a check to the credit card bank. Credit cards with "double cycle" billing should be avoided. If the credit card company uses this process and you pay almost the entire bill for one month, you may be hit with an interest charge the next month for all of the purchases from the prior month, even though you've already paid most of them.

You need to be cautious of "teaser rates" that tease you into applying for that credit card with an incredibly low APR. That rate can be increased at any time as long as the credit card bank gives you a fifteen-day notice. An alluring low introductory rate can climb as high as thirty percent or more. If this low rate you signed up for expires suddenly, you will end up paying more even on charges you already made.

Be wary of "double cycle", "dual cycle" or "two cycle" billing. This type of billing is a little confusing but allows you to avoid credit card charges only if you have paid the last two balances in full. For example, if you have a $1000 balance due in December, and pay off half of that balance, you'd expect to only be paying interest on the remaining $500 December balance in the next billing cycle. Not so with double-cycle billing! In January, when you pay off the entire balance, you'll also be charged interest on the entire $1000 December balance, not just the $500 that carried over.

Late payments can raise your interest phenomenally. Instead of the credit card bank rejecting charges that exceed your credit limit, they often let the charges go through but then charge a sizeable penalty as high as thirty-nine dollars. These fees add up and increase your balance and APR is also configured on these fees.

If penalties were not enough, you will pay more fees by paying by phone or using your credit card overseas. Oftentimes, you will not know about these additional fees until you're on the phone with the credit card company since they aren't in the disclosure materials or existing cardholder agreement.

Let's say you transferred a balance from one account with a higher APR to take advantage of a new one with a lower APR. This may come at a high cost. Any payments you make typically are applied first to the lowest rate balance. So while the credit-card company uses your payment to quickly pay off that 0% transfer balance, you are piling up interest on your new purchases, at say, 18%. Multiple balance transfers will also lower your credit score, which can raise your interest rate.

The Federal Reserve Board has released a final rule on credit card protection that will be effective beginning July 01, 2010. Unfair and deceptive practices by the credit card industry encouraged federal regulators to recommend a strict rule to give consumers a fair shake on their credit cards. The final form of this rule can be viewed online at creditcardreform.org/pdf/Final_Rule.pdf.

The new credit card rule will protect consumers by restricting increases in APR on existing balances, ensuring consumers have time to make a payment, ensuring payments are fairly divided among different balances, eliminating two-cycle billing, and restricting the financing of security deposits and issuance fees.

Now that you have read the preceding information regarding credit cards and how they can be detrimental to you, you need to learn how to manage your credit cards and get them paid off. April fifteenth seems to sneak up on us and many Americans have a tendency of either getting stressed out in filing our tax returns or are overly excited about the tax return we will be receiving and how we can go out and spend that money. This is the time to reprogram your thoughts.

Firstly, there is no reason to become stressed out because this is what is. Secondly, stop thinking about what you are going to go shopping for when you get your check. You need to realize that for the past year, you have been spending and spending and spending. For those of you who have no credit card debt, I commend you! For those of you who do have credit card debt, there is a way to become more financially stable and get some if not all of your credit card debt paid in full.

When your income tax return refund check or direct deposit arrives, stop. Look at your credit card debt and tally your balances. Depending on your total outstanding balances versus the refund you received, pay off your credit cards as soon as possible. If you are unable to completely payoff your debt, start paying off your highest APR credit cards first. Then move on to the next. By eliminating the higher APR credit card balances, you will save money in interest charges that are higher than your other cards.

The next thing to do is remove all of your credit cards from your wallet and look to see which card carries the lowest APR with no annual fee. Replace this credit card into your wallet and use it for "emergencies" only. That does not mean run out and buy a new big screen television because your old CRT television isn't working quite the way you think it should. That is not an emergency; even if your television doesn't work well, you could live without it for a while until you physically saved the money to replace it.

If you have sizable credit card debt, like I just said, pay down your higher APR cards first. You may need extra time to get them paid in full, but you can get there. Do not make any new charges on your credit cards unless it is absolutely necessary. If you receive a bonus from your employer during the year, use the same principal in applying the money to your credit cards to get them paid off as soon as you can. This process could take several income tax refunds and or bonuses to get the desired effect, but you will be much happier when you reach the point of not owing your soul to the credit card companies.

Credit card companies want you to charge when you make a purchase, because they realize that you, like many Americans, will not pay the balance in full when you receive your statement. That is how the credit card companies make their money with the ridiculous fees and APR that they set and demand.

I personally used to play the credit card and Uncle Sam game. I would charge up my credit cards all year, make minimum payments and when my tax return came, I paid off my balances. Then, I started all over again not learning from my own mistakes. Now, if my husband and I want to make a purchase, but do not have the cash on hand to do so, but have the money in the bank, we will use our credit card to make the purchase. When the credit card statement arrives, the balance gets paid in full. We do not charge what we cannot pay off at the end of the month.

I understand because I am a parent that circumstantial situations (medical bills, emergency dental work, etc) do arise in our lives with ourselves or our children when we do not have a choice on having to pay for something with a credit card and not being able to pay the balance in full when the statement arrives. When this does happen, do not use your card again until you get your balance paid in full.

If you have trouble managing your credit cards and continually find yourself in a situation where you feel like you can never get out of debt, once you get your credit card balances paid, start closing your credit card accounts until you are down to one or two. Or contact you credit card bank and ask them to lower your credit card limit. Do you really want a two thousand dollar limit on your credit card for a department store when you don't really need a limit that high for things that are unnecessary? Lower your limits to a couple hundred dollars; that's all you really need if you really need it at all.

Once you are in a situation where you can pick and choose the best credit cards for your life, consider "rewards" credit cards with no annual fee that offer a rebate returned to you, mileage awards, etc. consider making the change and replacing your old accounts with just one credit card that pays to use it. Just remember it needs to be paid in full each month so you can save up to hundreds of dollars in relentless APR and fees. Put that money in your pocket or put it towards a different debt that needs to be paid off. A really good option would be to send additional money in with your payment to your mortgage company to reduce the principal owed.

There is a lot of information everywhere on television, radio, print media and the Internet on ways to pay off your debt. If my suggestions will not work for you, do some research and find something that will work. Do not pay a credit card counseling agency or some other company to do the same thing you can do for yourself. Don't waste the money. The best strategy is to devise a plan that will work for you and use your head to make the best possible choices. Once you have cleaned up your debt, take a step forward, move on and do not look back. You just proved that you can do it, but is it something you really want to ever do again? I didn't think so.

# Chapter 7

## I don't want to cook tonight

We all love to eat out at some time or another. But many of us eat out more often than others. Several years ago, my family and I ate out sometimes as many as four nights per week and that was not including stopping by McDonald's, Burger King, or Wendy's for lunch. At the end of the month we could not believe how much money we spent on food outside of the home. I personally have always preferred eating at home due to the fact of working for a restaurant chain for twenty plus years. I prefer home cooked meals instead of the greasy cost consuming food in public. Take a look at your current spending habits and what your weekly or monthly cost comes out to be. Next look at how you can curb some of these trips in which you spend unnecessary money. It takes willpower to stop eating out as much, but if you have the *"desire"* and *"motivation"* to save some of that needless spending it can happen. Currently, my family and I do not usually eat out more than once per week unless we stop by McDonald's for something quick. We have learned to save a bundle each month by cutting back and we also feel healthier because we are not ingesting all of those greasy and fatty meals.

If you are in a hurry and cannot bypass stopping at fast food for something to eat, try to keep your budget in mind. Stopping at a convenience store can be healthier than fast food if you must stop. An apple or a sandwich is much healthier than a greasy burger. This trip is only going to be to tide you over until you can eat something more substantial after you get to where you are going. You drive into the drive thru line and see the familiar menu board with all of your favorite combos on it when you realize that you just can't cut back and you will make a better effort next time. Stop! You just made the most common mistake that these fast food places want you to make, you order the combo and then decide to super size it. Are you really stopping for greasy food that will help raise your cholesterol level and to gain an extra pound or two from these continual trips? This is where you need to change your thought process and recondition your brain. The bottom line is money, what you can save but yet get something to satisfy your hunger and taste buds. Have you ever noticed that the combo meals are directly within your view when you look at the menu board and that you have to search for the dollar menu and side items? This is actually done strategically by the company that designs these boards for drive thru fast food. The next time you pull into McDonald's, Burger King or fast food chain of your choice, stop for a minute and really look at their menu. (If you have time, go inside and look at their menu board at your leisure until you have understood what I am trying to explain to you) Now that you have had the opportunity to look, go back again and look at only the dollar menu. Remember that this trip through fast food is only something to tide you over until later. Pick out a dollar sandwich and a small drink or pick a small order of fries and a small drink. Your cost will be two dollars plus whatever sales tax that your state requires. You do not need (you may want) to eat a huge combo meal that in the long run is not healthy for you. This was just our first trip for the month. Now let's say you stop for lunch five days a week, four times a month, do you know how much you have just spent on eating lunch on the run? Do the math; forty dollars plus sales tax was spent frivolously that could have been spent on something much more necessary or put towards a household bill. Your next step is to figure out what you spend eating out for dinner. Add that to your lunch fast food total. That is a lot of money that many of us do not even realize that we are spending because we have to eat. Are you in a position to possibly miss paying one of your bills on time or not having the money to purchase a necessary health and beauty item for your family? I think you might get the picture. Just remember that there are alternatives to eating out and many discounts that we as consumers can utilize to save us money.

We have already cover grocery shopping and how to save money, but now is the time to budget into your grocery list some items that you can purchase to save you money at lunch time. Purchase an additional loaf of bread, some lunchmeat or spreads, some fruits and crackers. In the long run, you will be eating better quality foods and spend less money overall. Another item you could take for lunch might be leftovers from dinner the night before. Save where you can and eliminate the process of wasting food and money.

You really want to eat out because you had a rough day, you don't want to do dishes or because you just want to. Let's go back to the computer and start some research. Many restaurants currently (while I am writing this) have E-clubs in which you sign up for their newsletters, coupons and specials. Culver's (culvers.com) offers a buy one combo and get one free, Old Country Buffet (oldcountrybuffet.com) offers buy one buffet and get one free. Fuddruckers (fuddruckers.com) offers a coupon for five dollars off and Red Robin (redrobin.com) offers a coupon for three dollars off of a purchase and a free burger on your birthday. Wendy's, (wendys.com), Denny's (dennys.com), Famous Dave's, (famousdaves.com), Quizno's (quiznos.com) and other local restaurants in most areas offer E-clubs with email sign-up to receive these great discounts. Each person in the household should sign up for the same E-clubs so that you can utilize the coupons to save you money in the event that four people may be going out to eat. With two coupons and four people, you have the opportunity to save a bundle if your *"desire"* to eat out is stronger than your *"motivation"*. Many of these companies offer birthday and anniversary discounts as well, so you can save money on a special day.

The Internet is filled with so much information that it would take us lifetimes to read and review it all. This gives you a basis for free research and unlimited information and different scenarios on saving money while eating out as well as different areas. A well known website restaurant.com gives you the opportunity to purchase restaurant certificates at discount prices. There are some great opportunities available if you watch and wait for them. Occasionally, you can purchase a certificate for a certain restaurant for two to three dollars for a twenty-five dollar off of a thirty-five dollar purchase. The out of pocket expense on this certificate would run about twelve to thirteen dollars (two to three dollar initial investment and the difference in savings at the restaurant of ten dollars). It would be very wise to check with the restaurant personally by telephone to be certain that they accept this certificate. In certain circumstances, the restaurant may drop out of the program and restaurant.com continues to offer certificates for this merchant. If you are a frequent flyer with United Airlines, their website gives you an opportunity to use your miles to purchase certificates at restaurant.com. This is a great alternative to saving money when you feel it is necessary to eat out. Please consider reading all of the rules and regulations associated with the certificate from different merchants as they may have restrictions or black out days.

So, you still don't want to cook tonight. If families shared more interaction together in the home, I believe that families would have a stronger bond. Let's see if we can strengthen that bond here in the kitchen. If you have school aged children who want to participate in helping with dinner but you feel they are just in the way, let's change your thinking a little and give them a minor chore to do. Remember that eventually, your children will be old enough to function in the kitchen without the help or supervision from an adult. Once they are old enough, make a night for family night cooking in which each family member helps prepare the meal. In the summertime, grilling is quite popular and can be a lot of fun. If you have a grill with a propane burner on it, the entire meal can be prepared outside. The children may not seem overly enthusiastic as they get older in helping prepare or even preparing an entire meal, but the results will be wonderful. Not only have you taught you child to cook, but to be integrated in a family that does things together to teaching them that it is almost always cheaper to eat at home than it is to eat in public.

One of the easiest ways to stay within your budget is to have a budget. Money spent purchasing groceries for eating at home, is money well spent. There is no comparison at all for eating out versus eating at home. The cost to eat out on a regular basis could very easily double if not triple what you would spend on your grocery bill. A wise decision in getting started would be to get a pen and paper. Sit down and think of only the necessities you require to replenish what you are out of. Tack this list on the refrigerator and each time you are out of a staple write it down. Now it is time to go shopping and you will choose four items for snacks for the next two weeks. Try this several times when it is time to shop and eventually you will catch on and also see a difference in the savings at the end of your visit to the grocery store. Do not go to the grocery store hungry or you can blow your budget with unnecessary items such as ice cream and snacks. Your budget is first and your stomach is second!

# Chapter 8

### Is it time to move?

In one of the later chapters in this book, I discuss the pros and cons about rent vs. buying. There is plenty of information regarding the choices you face if you are considering purchasing a home. This chapter is not related to purchasing a home, but will cover the situation of renting and the consideration of moving to another rental.

Many people have a tendency of moving from one apartment property to another as soon as the lease on their current apartment expires. There are many things to reflect on when this situation arises.

Location... location... location. We hear this many times, but usually relate it to business and think about whether a business will prosper in a certain location. In this instance, I relate location to convenience of your lifestyle. It is important to be centrally located in where you live in relation to job, school, family, etc. Is it cost effective to live thirty or forty miles from where you are working? Drive time to and from your place of employment can and will take valuable time out of your day. The excessive miles on your vehicle and the extra cost of fuel can easily impact your budget and deplete funds for entertainment with friends and family.

Your rental should be cost effective for your budget. It is unnecessary to spend money on a rental that is beyond your budget. Living within your means will produce a budget that is easier to manage. Stretching your budget by overpaying on rent will usually infuriate you and cause you stress if something financially negative comes up. A renter should not spend more than one third of their monthly income on rent. If it is possible to spend less, then do so. This will give you more savings power later on down the road.

You may currently live in the perfect rental, but are considering a move because your lease is due to increase by twenty-five dollars a month. You are furious and don't feel that you should pay the increase when the complex down the street is offering a great move in special of $100 off your first three months of rent. The rent would be the same as to what you would pay when your rent goes up twenty-five dollars. It is time to take a deep breath, relax and sit down to do some math. The first step is to realize that you would only save three hundred dollars from the new apartment over a one-year period by moving. The second step is to see that you would spend three hundred dollars more per year if you stayed where you are currently living.

There are many things to take into consideration when you get ready to move. The transfer of utilities can be added into the cost of the move. Some companies charge a reconnect fee or a transfer fee which may not sound like much, but will add up with the other expenses of moving. You may need to rent a truck to move your larger items along with stacks and stacks of boxes you packed. This expense can be costly and run over a hundred dollars. You may need to have friends and family come help you move, which may seem free at first until you buy pizza and soda for everyone. Another expense would be reordering checks from the bank for your new address with the cost anywhere up to twenty-five dollars or more. (If you switch to generic checks, it will save you money. No one really cares what designs are on your checks)

Re-evaluating your two segments here is important. Is it really cost effective to save twenty-five or thirty dollars moving to a different property? Over the one-year period you may spend that just in fuel going to the same places you were going to before your move. The grocery store, gas station, your job, etc may only be a few miles further than you were driving before, but the additional miles will add up along with the stress and frustration in making this move happen.

If you need to move due to employment or school reasons, you may not have a choice in spending money to relocate. Be smart about your choices and keep your monthly budget in mind. Consider your income, your necessary expenses and your housing expense. It is wise to keep your rent payments below thirty percent of your gross income, but is wiser yet to spend twenty-five percent or less per month.

With the crisis that our economy is currently in, there is another option for you to keep your money in your pocket that you would have spent on a move and still save on your rent without moving. There are many slow rental markets across the country and many properties are not renting at full capacity with numerous vacant units. It is not unthinkable to approach your property manager and request to negotiate the increase on your renewable lease. Many times a compromise can be made and the possibility of your property manager matching another property's special. This may be implemented for resident retention. If you move, that is one less unit the complex is receiving rent for and could mean a loss of eight thousand dollars or more per year for them.

Always look at the big picture when it is time to make this type of decision. It could be very costly on your part if you do not weigh the pros and cons to moving or staying where you are. It is very important to not lose sight of your goals

# Chapter 9

### Talk is cheap, but is the phone?

Most Americans have at least one cell phone if not two. Cell phone users have all benefited from the convenience of on the go phone usage and the experience of no longer needing to race home to receive important messages on your home phone. Most cell phone plans are similar for monthly minutes and monthly rates. There is one thing that I personally would not skimp on is the text messaging plan. Although my carrier is Verizon Wireless, I know that other carriers offer similar texting plans. I pay a twenty-dollar a month fee for unlimited texting. (I know some carriers only offer up to a certain amount and the rates increase by how many messages are sent.) With having two children below the age of twenty-two on our plan, the texting volume is unbelievable. If I had to pay for each text message coming in and going out, I would not be able to afford it. With having the texting plan, this gives me the opportunity to lower my monthly minutes (the kids prefer texting over talking on the phone and with our carrier we get free mobile to mobile within Verizon Wireless customers.) I also keep the kids on our plan even though one no longer lives at home. It only cost an additional ten dollars a month for him to be on our plan versus sixty to seventy dollars per month to be on his own. Once the kids are over the age of eighteen, they purchase the phone that they want (they have a tendency to pick one of the newest phones that do everything but wash your clothes and the newer phones are more likely to be much more expensive than what is necessary).

Recently, my husband's contract was coming due where he was eligible for a discounted phone with a new two-year contract. We drove to the Verizon Wireless (VZW) store to see what options that he had. He did have to take into account that his job caused him to be a little bit harder on his phone and to consider durability. He looked and tried the phones and observed the prices of them but was not satisfied with how much he would have to pay for a phone he really was unsure about. We left undecided because in his view eighty dollars was far too much money to pay for a phone. Once at home, he sat on the couch and began reviewing the ads in the newspaper. Low and behold, he found an exact model cell phone at Walmart for twenty-eight dollars that we had seen at VZW for eighty dollars after rebate. My husband was speechless for a minute. He called over to the VZW store and spoke to the salesperson that he had been working with when we stopped by earlier. He explained everything about the ad and asked if the store did any ad matching; we were told that they do not match prices and were sorry. I had a brainstorm right after that and thought about how many times customer service at VZW always made things better when I had a problem. I figured that this constituted as a problem; I picked up the phone and held my breath with the hopes that someone would be on my side. A very nice female representative took my call and listened to my woes as I explained to her what had transpired over the last sixty minutes and the annoyance my husband now felt. She asked me to hold for a few minutes and came back with great news; her manager approved us for purchasing the same phone that Walmart advertised (the same one we looked at, at the VZW store) but for thirty dollars after rebate. We received the cell phone four days later and my husband was happy. When dealing with merchants, sometimes it pays to go to the corporate level to save money. In our case, we saved fifty-dollars!

# Chapter 10

**Rent vs. Buy**

This is a great time to buy a home. The interest rates are at an all time low since the 1950's and home prices are lower than they have been in years. There is lot of inventory on the market due to the rate of short sales and foreclosures. It is a great time to rent. Many rental properties are offering great move-in specials. What do you do and how do you make the choice on renting versus buying? Is it better to stay in your apartment or buy a house or is it better to sell your house and move back into an apartment?

At the present time, you may be renting an apartment/living space and are considering the purchase of a new home. Renting an apartment may suggest hassle free living and there are several pro's to renting in such that you can come and go without the worries of homeownership. It is usually more convenient having a maintenance-free environment without the concerns of repairing or replacing items in a house. This can free up many hours a week. Apartments are usually smaller than the homes that we purchase, which will mean less time cleaning our surroundings in an apartment. Many Americans find that there are some cons to renting living space such as dealing with non superior property management and or maintenance who disregard your concerns or needs, noisy neighbors that affect sleeping habits of your spouse or children, other residents parking in your assigned parking space, residents who are uncaring about debris they have dropped or left in unauthorized areas, pet nuisances, etc.

Homeownership can be a blessing or an overall disaster to someone's life. It is definitely nice coming home, driving right into your empty garage and going into your nice quiet home. No one lives above or below you and you get no complaints that you are making too much noise when you turn your stereo or television up a little louder. The cons to homeownership can be overwhelming if you are not prepared for them. The cost of having to repair or replace appliances can be costly and so can those remodeling projects. Security is another issue because you have to devise a plan that works best for you. A lot of homes that are purchased do not have any type of security or additional lighting to make you feel secure in your new neighborhood. Suddenly, you are slapped with the proposal of should I spend hundreds if not thousands of dollars upgrading the security for family. Many first-time homebuyers do not realize the increase of utility usage that a house requires over an apartment or even the cost of trash removal or water usage.

Renting a house or apartment can be costly and add up over the years and then you have nothing to show for your money at the end of thirty years when you could have owned a house. Depending on what you spend in monthly rent, the appreciation of rent and rental insurance could definitely be much greater than owning a house. Remember that buying is not always the best solution to renting. Here are a few examples of good versus bad scenarios for buy versus rent. The examples are based on a thirty-year time frame at a fixed interest rate. A renter pays $900 per month in rent ($438,135 after 30 years including inflation) and for a $200,000 house with 3.5% down, yearly property taxes of $2200, yearly homeowner's insurance of $900, interest of 5.5%, ($520,261 after 30 years) With tax/ interest benefits you would save $67,846; even with appreciation, but you would save $60,836 by renting. A second scenario is where the renter is paying $1000 per month in rent ($486,817 after 30 years including inflation) and for a $150,000 house with 3.5% down, yearly property taxes of $2200, yearly homeowner's insurance of $900, interest of 5.5% ($401,636 after 30 years) With tax benefits you would save $55,009; even with appreciation, you would save $30,788 by owning.

The two above scenarios do not include any maintenance repairs or replacements for owning a house (roof replacement, new siding, new kitchen or bath, new appliances, etc). You will need to determine what you think you may need to replace on a house you buy and guesstimate if the expenses will exceed the savings of owning versus renting.

Before you decide to buy a home, sit down and do your homework before you contact a real estate company. An agent will discuss affordability of home buying, but many will not suggest that you look at the big picture of all that is actually included. The very first question you need to answer is %How much money do I/we have on hand for a down payment?%If you have no extra money and are not a veteran, it is most likely best for you to wait in purchasing a home. Start saving your money diligently and reassess your finances in six months to a year. There are a lot of unexpected fees that you do not expect when purchasing. To purchase a home in today%s market the general rule of thumb is 3.5% down and closing costs. Homeowner%s insurance and pro-rated taxes are not usually financeable and will need to be paid at the closing table. It can be quite a surprise if you are working with an inexperienced Realtor® or loan officer.

If you do have money in savings, you will need to formulate a list of all expenses in purchasing from moving expenses, cleaning expenses, utility hookup, phone service transfer, reordering checks, etc. If you are unsure of what average gas and/or electric bills run, ask your friends or family who are homeowners. They can give you a general idea. If there is a particular home that you are interested in, most if not all of the utility companies will give you an average of usage over the last year and will even give you a high and a low reading. Everyone consumes energy differently and it also depends on the size of the house, the insulation and HVAC system. These are your start up costs just for moving.

Once you have come to an average expense of these items, start a new list that would consist of monthly bills. This may include but is not limited to utility expenses, trash pickup, water service, phone service (land line and cellular), cable/satellite service, auto insurance, credit card payments, loan payments, etc. Now, it is time to add that up and determine what you can comfortably afford to pay as a mortgage payment.

If this is within your budget and does not put you on the edge, it is time to shop around for a mortgage company to see if you qualify for a loan and for how much. A mortgage broker can give you a guesstimate on what a payment will run for a certain dollar amount at the current interest rates.

So, you qualify for purchasing a home! This is one of the biggest purchases you will make in your lifetime and this decision deserves a lot of thought before you come to a final conclusion. Also, if you are planning on purchasing a house with your spouse or significant other, remember that this is a joint decision and it is very important that you both have complete input and are in full agreement on the choices that you are about to make. If you are not, it could be financially devastating if you needed to sell your house in the first couple of years of buying it. On a $200,000 house that is purchased at 5.5% interest, you will pay a little over $10,000 in principal in three years and if you have to sell it due to a poor decision, you will have a commission to pay your real estate (the cost is 5% to 8% of the sale price) and closing costs plus your initial investment. The chances of breaking even are slim unless the market has appreciated so intensely, which I hardly doubt with the economy the way that it is. So, make wise, not hasty choices.

If you find that buying a home is not in your budget because of credit, finances or the unwillingness to spend on what America calls an investment, don't worry about it. The time will come when you decide to take the journey through home ownership when the time is right for you or if you ever feel it is right. I personally, would prefer to rent as it is cheaper in the long run and is less time consuming.

Here is something to ponder. Life is not keeping up with the Joneses, not having everything that we want or ending up owing more than what we can ever repay, but what you make it. Enjoy yourselves, your children, your family and friends. Enjoy your freedom (that is why we live in America) and enjoy what time we have been given to live on earth. You will find that you will be happier and not as stressed if you maintain a healthy financial lifestyle and not worry about what everyone else is doing or spending. What works best for you, may not work for someone else and vice versa.

A different scenario could be where you currently own a home and you're in a situation where your monthly mortgage payments are higher than what you would like them to be. Depending on what the interest rates are, you may be in a position to refinance your current mortgage. However, if you choose this route, there are some things you will want to consider. What type of loan is the mortgage company suggesting? How many years will you be financing? How much is the bank or mortgage company charging you to originate a new loan and what will be the closing costs? The next question is how many more years are you planning on staying in the house? If you will only be staying in the home for a short time, it is not worth refinancing your mortgage. If it is long-term, meaning at least 7 to 10 years, it could be worth refinancing if the rates and fees are good.

If your decision to stay in the home is short-term, it is best to continue making the current monthly payments on your loan until you are ready to relocate to another place to live. At that point, it could be advantageous for you to review the information preceding these passages regarding rent versus buying. If your payments are higher than what you would like and the interest rates are no longer at their all-time low, it could be a wise decision for you to go back into renting. This could be a cost saving measure towards your finances and give you the opportunity to save some extra money so you are in a position to be able to do some of the things you would really like to do.

Many people are very happy renting an apartment and many people are very happy owning their own home. This is a decision that you will need to make for yourself that no one else can make for you. This is an area where you and/or your spouse or significant other needs to choose what the best decision is for your own lives. Owning a home can be a big responsibility but can be very enjoyable for the right people. Whereas, renting can free up some of your time everyday and lessen some of the load financially. Whatever decision you make, needs to be made for you/spouse and not for your family and friends.

# Chapter 11

## Chop chop...bye bye mortgage?

So, you have decided that it is time to consider paying off your home mortgage early or eliminating it altogether. You could be in a position financially where this could be very easy for you or in a position where you are financially strapped. Regardless of either situation, as long as there is "desire" and "motivation" it can be accomplished.

Many people believe that they need to refinance their current mortgage into a 15-year mortgage to pay off early. This is not the case. There is a lot of media and advertisement giving information about refinancing your home and many people do not realize that they can basically do a "refinance"on their own without even needing to talk to anyone. Refinancing your home loan will cost you money. Even though the bank or mortgage company will tell you there is no out-of-pocket cost, there is because you do have to take money out of your pocket to pay your mortgage every month. Some establishments advertise and offer no closing costs for refinance, which in turn causes the average consumer to believe that they are getting a loan with no additional fees. This is not true. Why would they offer to do something like this for free and not make any commission on it? Because, they do not do it for free. Many consumers are not aware until they sit down at the closing table to sign the papers for the refinance that there is a fee on the good faith estimate for loan origination. This charge will usually cost you 1% of the total amount that you will be financing with that establishment. Depending on how much money you finance could mean several thousands of dollars added to your final total. There may also be other fees incurred and it may seem minor, but once added into your total it could be somewhat devastating or depressing to see what you have just done.

The easiest way to refinance your mortgage and save the most money will be on your own. If you are in a position financially, all it takes is to send an extra $200-$300 more to each payment per month. (If you could send one and a half times your mortgage payment to your mortgage company, this would be ideal to pay off your mortgage in a much lesser amount of time.) Whether you use the Internet or you receive a statement in the mail, there is a place where you can allocate this money to go to. This extra money needs to be paid on the principal and not on escrow. If you were to make a $300 payment towards principal in addition to your regular monthly payment, you would eliminate interest on $3600 per year. It may not seem like much, but it will add up quickly. In three years, you will have paid an additional $10,800 that you did not have to pay any fees on.

If you are in a situation where there is not very much money left at in the month, but you can afford an additional $50-$100 per month add it to your monthly payment and allocate it towards principal. In three years, you will have paid an extra six hundred to twelve hundred dollars without having paid interest on that amount.

Only you and/or your spouse/significant other can make this decision. It is best to review your finances before making a large principal payment on your loan. If your household receives an employment bonus anytime during the year, this money can also be sent to your mortgage company and put towards the principal. It is better to send additional money every month, than to send a small one lump sum once or twice a year.

Another option that some mortgage companies offer is to make your payment in two different parts. If your payment is due on the first of the month, you will send half of your mortgage payment on the 15th of the month and then pay the other half on the first of the month. This will reduce some of the interest that you are paying on your loan. It may not seem like a lot, but this amount will add up and save you hundreds of dollars over the years.

To completely eliminate your mortgage loan long before the note is due, you will need cash to pay it off. There are no miracles like winning the lottery or getting some form of inheritance to pay off your debts. It does happen to some, but the chance of it happening to the average consumer is slim –you have a better chance of getting struck by lightning than coming into a large cash flow. If you need to completely eliminate your mortgage for financial reasons, it is best to find the right option in selling your home or to rent it out instead. Keep in mind that if you do decide to sell your home, you do not need to be hasty in choosing how you will do it. There are a couple of different options for that which would include finding your own buyer and having an attorney draw up the paperwork for the sale of your property. There are also companies that will facilitate the sale of your property through a "for sale by owner" company. Or you could contact a local real estate company and interview not just one but several real estate agents and find out who will do the best possible job in advertising to get your property sold in a timely manner. Remember, you get what you pay for but commissions are negotiable and are not set in stone until you sign the representation from the agency that will sell and market your house.

You will need to hire a real estate professional to sell your home if this is the choice you have made in eliminating home ownership. It is very important to know what services your Realtor® will provide for you when you have them list your home. It is always best to interview three or four agents before deciding whom you will use. In the event that you have a friend or family member who recommends an agent, interview them as well. Just because someone has indicated that they did a great job for their transaction does not necessarily mean that they have all of the tools and experience to sell your home. Research is imperative because you will pay a hefty commission regardless of whether an agent has fulfilled their obligation or not. Do not be afraid to ask for references. If you are not satisfied with the only two or three provided, ask for more or move on to another agent. You want to be able to pick and choose which references to call and have a list provided of many to choose from.

Get your money's worth. An agent should inform you during the listing presentation on what they are willing to do and try if they get your listing. Advertising and marketing are important as well as where they can find prospective buyers who are looking for a house like yours. Will your agent hold "open houses" and how will they advertise and market your home? Will the agent lower their commission? How will they advertise and how often? How will they find prospective buyers? What will be the game plan in getting your home sold for the best possible dollar?

You should have a good relationship and communication with your agent. Your agent should be in constant contact with you regarding your listing. After you home has been on the market for thirty days, your agent should be coming to you to discuss a price reduction on the property. If your home has had no offers but a lot of showings, you may need to make some home improvements or take that price reduction to get your home sold. Do not hesitate to ask your agent for proof that they are actually advertising and marketing your home. There are many websites on the Internet that have property listings. Ask your agent which sites they advertise on and if they have a personal website that advertise their listings.

# Chapter 12

### Teaching your kids about money and how it really works

Children are like sponges and learn best about finances as soon as they show an interest. That does not mean your three-year old is going to ask you how you pay the bills. By the time your child is three, there will be some form of interest in money. They will understand that money will buy something they want, but not the part about how much money it will cost. This is a good time to sit down with your child and begin the process of teaching them about the difference between, quarters, dimes, nickels, and pennies. If there is not enough interest wait six months and try again and so on. By the time your child reaches kindergarten, they should understand the basic concept of coins and paper money. Each child learns at a different rate; you will know when they are ready. Do not force your child or they may show resentment and intentionally to not grasp the idea.

When one of my children was in the third grade, he had difficulty learning fractions. He was irritable and stressed over learning this new concept. One day, I had a brilliant idea that helped him get through this difficult time and lead him to better future understanding of money and mathematics, which he came to love. I gathered all of my coins and had him sit at the table with me. I spread out four quarters in front of him and asked him what he had. He told me that he had one dollar and I agreed with him. I then removed two quarters and asked him what he had and he told me he had fifty cents. I explained to him that although he had fifty cents he also had a half of one dollar. He understood, so we continued on to one quarter and three quarters. We worked on the project of quarters for about thirty minutes. On the next day, we graduated to dimes and so forth down the line until he grasped the idea of fractions and also learned the value of the coins better than he had before we had started.

Remembering back to when one of my children was about seven or eight-years old, I recall how he would always leave his spending money at home whenever we went to one of his favorite department stores. He would see something in the toy department that he really wanted and would then beg me to lend him the money to buy it. He was always willing to pay me when he got home. Over time, I became frazzled over the money he threw away on toys that he would only play with once or twice. One day, I decided to try a different approach to his begging. I became firm and decided to not give in. The next time we went to the store, he played his usual antics and asked to borrow money until we got home. He was devastated when I told him that I would not lend him the money this time. I explained to him that next time we came to the store that he could bring his money with him and if he still wanted the toy he could buy it then. I played this game several times until my son caught on. On a following visit, he would ask again to borrow money for a different toy and I would ask him what happened to the toy he wanted the last time we came. He told me that he had changed his mind. After four or five times of doing this, he realized that the toys that he had the impulse to buy were not the toys that he really wanted. I had changed his thought process into thinking first before spending.

Each child and each person is different. Spending may be difficult for some while saving money is difficult for others. Some children feel the need to spend that quarter or dollar in their pocket before it burns a hole in it, while others hoard their quarters and dollars in jars and piggy banks. If you start while your child is young, you will have a better chance of teaching them better fundamentals of saving their money for special occasions.

Out of my three children, my youngest is very frugal with his money. He is old enough to drive and already has his own used car that we purchased for him. But he chooses to be cautious about how much he drives because of the upkeep on his car and the cost of fuel. Most teenagers are beyond excited for taking the car out with their friends and cruising around town without the thought of draining the gas tank.

While growing up, my middle child had the burning hole in his pocket syndrome. When he wanted something, he wanted it regardless of whether he would use it more than once. I think it was the novelty of having something new or something his friends didn't have. Now that he is on his own and responsible for most of the facets in his life, he has become quite frugal in spending his money. He does not eat at home which gives off the impression of bachelorhood, but he watches what he spends and how much he spends. He has come to the realization that having money in the bank is a good thing and makes him happy to have it.

My daughter knows all of the concepts of finances and how to keep a budget and also how to make it work, but has difficulties maintaining a suitable budget. She has the American Dream in her head, keeping up with the Joneses and status quo. As I have said before and will always say, regardless of how much time you have spent teaching your child the good fundamentals in life, every child is different. They will choose their own path and utilize what they feel is important to them. All you can do is your best when teaching them the important necessary tools to get them through life. Once they have developed habits of their own it is often too late to navigate them back to where they need to be. So, remember to guide them and teach them as early as possible.

# Chapter 13

## Is it time to fly the nest?

So, you think you are ready to make it on your own! How exciting! You can hardly wait to get out there and set up your own home and move away from your parents and their rules. This is a very electrifying time for you. Don't let me burst your bubble, but....yes there are always buts. You need to plan, prepare and put a lot of thought into this before taking the giant step. You will be able to make it on your own if you put all the "how to's" to work for you. It is not as easy as just going out and renting an apartment and moving.

Moving to your own apartment after living at home with family can be a little overwhelming. It may be hard to juggle all of the aspects when you were unsure as to what you thought you would be spending versus what your current monthly obligations are.

It is important to check into every feature and scenario before making the big jump to what you may call "freedom". Freedom has a price tag on it and if you are not cautious, you can become debt ridden in a short period of time. You can also ruin your "renting status" if you have to move out early and are unable to finishing paying your lease. By not paying your monthly utility bills associated with your lease can cause you to lose future chance of retaining service in the future or having to pay a hefty deposit.

There are many preparatory things to consider before you commit yourself to moving into an apartment for the first time. You should sit down and make a list of everything that is needed and things that need to be done before signing a lease. There will be a lot of questions you will ask and it is important to take notes.

You may think that you are financially ready and capable to handle this at this particular time in your life. But you may not be. If you are, I wish you the best in settling into your new place.

The biggest thing to consider is if you can financially take this step right now. And rightfully you may have the resources to do so at present. For those of you who are not quite so sure, continue reading so you can start the preparation mode to get you started on the road to "freedom" of living your life without the rules of living at home. You will need to ask yourself some of the following questions.

Do you have a steady income to meet your new monthly obligations? What is your take home pay per month? What are you current monthly obligations that you will be responsible for after moving? What can you comfortably afford to pay in rent each month?

What utility bills will you be responsible for? Gas, Electric, Water, Heat, Local phone, Cellular phone, Cable, Internet, etc.

How much food do you eat? What groceries an staples will you need? Do you nee to purchase snacks and beverages? How much money will you need to eat out?

What bathroom supplies do you need monthly? Shampoo & conditioner, body soap (liquid or bar) & body talc, Shaving cream & razors & cotton balls, hand soap, female toiletries (make-up, lotions, perfumes, monthly items), first aid items (bandages, gauze, antiseptic, ointments), toothpaste & mouthwash, etc.

Items you will need when you get sick (basic). Cough syrup, cold tablets, allergy preventers, headache relievers, stomach aids, etc.

Paper products & cleaning supplies you will need (basic). Toilet paper, paper towels, window/glass cleaning, tub/shower cleaner, toilet bowl cleaner, plunger, air freshener, Kleenex, toilet bowl scrubber, etc.

What household supplies do you need? (basic) Cleaning supplies, dish soap, dishwasher soap, floor cleaner, glass cleaner, surface cleaner, bathroom cleaners, sponges, scrubbers, wash towels, oven cleaner, dust cleaner/polish, paper towels & rags, cleaning bucket & gloves, vacuum cleaner, mop, dust pan & broom, dishes & silverware, glasses/tumblers, pots and pans, baking ware, food storage containers, foil wrap, plastic wrap, plastic food storage bags, etc.

Miscellaneous items will be needed. Salt & pepper shakers, can opener, serving utensils, cooking utensils, colander, dish towels, pot holders, mixing bowls, bath towels, etc

Items you will need for the living room. Couch, loveseat, chairs, end tables, entertainment center, shelving, television, lamps, etc.

Items you will need for the kitchen. Toaster, microwave, blender, coffee maker, hand mixer, clock, canisters, silverware and separator, dishes, pots and pans, bake ware, dining set, etc.

Items you will need in the bedroom. Bed, dresser, might stand, shelving, hangers, lamp, alarm clock, etc.

Groceries are necessary. Meats, veggies, fruits, dairy items, packaged items (pasta and rice), jarred items (peanut butter and sauces), condiments, beverages, snacks, canned goods, bakery items, etc.

Do you plan on living alone or planning on having a roommate? Do you have a roommate in mind? Is your potential roommate financially stable with a steady income? Is your potential roommate trustworthy and reliable? Is your potential roommate similar to you? (In the sense that if you are quiet, are they quiet? If you are active, are they active? If you are a partier, are they a partier?) (This can cause conflict between the two of you if you are opposites in the way you live) Is your potential roommate committed to this obligation?

Do you know what your financial obligations will be? RENT, student loans, auto loan, auto insurance, cell phone, Utilities (water, gas, electric, cable local phone), credit card payment(s), car repair allowance, gasoline, groceries, clothing allowance, entertainment, etc.

Have you chosen a place to live that is convenient to job, friends and family? What is the distance to & from work? Will poor weather affect your free time in a negative way? What will your gasoline expense cost you monthly? How will the impact of the drive influence auto maintenance? How far away will your parents live? How close to your friends will you live? Are there things to do in the potentially new neighborhood? Have you considered the safety in the potentially new neighborhood? What is the crime rate? For robberies, gang violence, etc there are different Internet site to check out. (citydata.com), (homefair.com) (familywatchdog.com), and (neighborhoodscout.com) are a few.

How safe is the apartment complex or building? Does the building/complex have a stairway accessible to your potential unit for emergency evacuation? Is there secured access to the complex/building? Are the grounds well lit at nighttime? How much do you know about the apartment community? How much does the credit check cost? How much money is needed for the security deposit? If you have a pet, is there a pet deposit? How much will rent cost? Does the property have a full time maintenance employee? What is the time frame for repair orders? What is the last day rent can be paid without a late fee? How much is the late fee? What are your obligations to the lease if you decide to purchase a house and break your lease? What amenities does the apartment community have?

If you were not aware of how much information was needed to lease an apartment the preceding information will help give you an idea of what to look for and what to keep in mind. There will be items in this list that you may not need and others that are not on this list.

The big question is how will you pay for everything you need? This is where this chapter will be helpful if you are trying to figure out how this can work out for you and not cause you to be broke before you even move into your new place. The following are just a couple of examples of how you can make this work for you and not cause you to become financially overwhelmed.

You teenager is interested in getting your first apartment and are "fried" with living at home with mom and dad. You cannot wait to venture out on your own with no rules or nagging. The idea to move out is so strong and the need to do so is beyond what your parents want. You have a job and believe that you are totally capable of paying monthly expenses without any concerns and will still have a little left over for emergencies. You need to set up a mock budget, which will include all prospective monthly bills and any unseen emergencies.

Unfortunately, there is not much money left over to get the furnishings and supplies that are needed to get the apartment ready for day-to-day living. There are several ways to acquire the things you need. Even though you think that mom and dad are nagging rule setters that will ruin your life, they can be very helpful in lending a helping hand. (Remember that they might be just as excited as you are in your new adventure. Parents like their alone time too.) Mom/Dad can check to see if they have any extra items that they can donate to them and even check with extended family or friends to see if they have anything for the new place. Someone may be able to come up with kitchen and bath items while others may come up with furniture and staples. Try visiting some garage sales for the basic items you think will be needed. The local dollar store is another good place to find inexpensive items that will come in handy at a fraction of the cost of a retail department store. It would be a good idea to start planning on finding donations several months before it is time to move. It can be very overwhelming to move into a new place and not have the items that are needed on a regular basis. It can be financially stressful to have to go out and purchase these items due to poor planning. It can cause anyone to become mentally drained and cause unhappiness with the decision that was made not long ago.

It is now time to move away from home and you will want to make the best choices to make it happen and for it to last. Many parents do not mind that their child is still living at home after graduation from high school/college and are genuinely interested in their independence of trying to make it on their own. You (the child) have a job and know that you are in a position to live on your own or are very close to making the jump. This is where the sacrificing comes into play for you. Since you are working and believe you can manage on your own, sit down with your parents and have a chat about it. You will need to figure out a dollar amount for rent that you can comfortably afford. (In this example, rent will be $800 per month for a one-bedroom apartment) Even though you have not made a commitment to leasing an apartment, you need to pretend that you will begin paying rent on the first of next month. Ask your parents to collect the rent from you for the next six months, but to put it into a savings account that you do not have access to. They have to understand that under no condition are they to return the money to you until you are ready to rent an apartment. At the end of six months, your parents will have helped you save $4800. (If you can save like this for nine months, you will have $7200 and $9600 after one year. You need to make the decision on what you need, when you need it, the best scenario for you and when you want to move). Many apartment communities charge an average of a $400 deposit and your first month rent. Depending on your credit history, they may require your last month's rent at the signing of the lease. (This would cost you $1200-$2000 depending on your situation.) If you are able to find an apartment for $800 per month, you could have $2800 -$3600 to get your apartment set up and still have money left over to put into the bank for reserve. If your plan is to purchase brand new with a few hand-me–downs, shop wisely. Discount departments stores and local dollar stores can save you quite a bit of money. The local newspaper advertising section or a community website (craigslist.org) can help assist in finding some good bargains on furniture. The local ads in Sunday's local newspaper can aid in locating the basic items that are on your list to get you started.

Keep in mind that this is only your first apartment and that you do not need to have everything brand new. Even now in my forties, I still have miss-matched items in my home as I am sure that some of your parents or friends parents have the same in their home. It takes a lot of money and most of the time many years to get things "perfect" in your home. And some people never achieve this because it may not be important to them to be "status quo".

If you do move out and have to move back home because it just didn't work out due to financial reasons or other circumstances, the time will come to try it again. The next time will give you the insight to prepare and plan for a smooth transition that should be beneficial to you as long as you remember that you need "desire" and motivation. "

# Chapter 14

### Is there a way to teach your kids to save for the future?

Teaching your children how to save money can be very difficult and other instances they may be quite resilient to the idea. Every child is different and some children may never grasp the idea. Just remember that the younger the child, the easier it is for them to learn and accept the idea as part of their life.

In an earlier chapter, I have discussed teaching your child to "back off" from spending their money and to put more thought into how they will spend it without impulse buying.

Teaching your kids to plan for their future is a key element in developing a stabile financial goal for them that they can see through to adulthood and even to retirement.

You child will eventually grow to the point that they may take on babysitting jobs or a newspaper route. This is where you need to develop this plan and follow it to the end. Once this plan is started, it may take some coaxing to keep it on the right path and to eventually have your child adhere to it willingly.

You will need to have this conversation with your child before the subject of earning his or her own "pin" money comes about. When my son was an older pre-teen, I sat him down and explained to him how things were going to "work" once he started making either money on the side or had a part-time job. I told him that because I was his parent and made the rules in our home, he had no choice but to abide by them. He heard me tell him that once he began earning money, half of that money would be put into a savings account that he would not be able to access. The money was to be used towards any extra college expenses that his father or I would not be able to cover. He was not overly excited about this, but he did understand completely the reason we would do this.

Next, I explained to him that once he started a full-time job after he was to graduate from college, he would need to put ten percent of his paycheck into the account he started years ago. I clarified the plan from there justifying that if he deposited this money every payday, he would have a sizable retirement plan when he got older. He totally understands that the Social Security Administration will be bankrupt and will not exist by the time he should retire in another 55 years. I also enlightened him to the fact that as long as he follows through with this goal, there would be a possibility of retiring much earlier than our US Government expects.

My son, who is already good with saving money and thinking ahead to future finances, posed a question to me and asked if there was an emergency or something big that he wanted or needed, could he make a withdrawal. I simplified this goal and indicated that with the knowledge that he has, he should not be in a position to convert this account to a spending account. He already knows how to manage his money to a certain degree and if he applies himself, he will manage. By managing his earnings with a full-time job after earning a college degree, he should be in a position to take additional money from each payday and put that into a usable savings account for when big-ticket items come up.

By withdrawing money from this off limits account, he understands that he will jeopardize himself and his future family from the security they need to depend on.

I have spoken with a professional friend of mine who had a client whose father implemented a very similar savings plan with him. He told my friend how well it had worked for him and that he was able to retire at a much younger age and still had extra money from this account to enjoy his "golden years" traveling to some of the places he had always wanted to visit.

This plan can work. It is not fool proof by any means. It will take a lot of hard work and determination from you and your child to get it off the ground and then stick to it. I would be financially secure sometime in the near future if not already had I implemented this goal into my own life at my son's age.

Many children do not pick up the newspaper to read what is going on in the world. They rely on the radio to hear bits and pieces of local and world news. While your child is still young and in middle school, begin with implementing the newspaper to them and have them read the local news. Eventually that can or will be expanded into reading world news. This is a good time to introduce them into reading about stocks. Have them choose a stock and take daily or weekly notes. This will give them the opportunity to see if their chosen stock is appreciating or depreciating in value and to give them a little insight into what the stock market is all about. One day when they have reached a point in their life, they may want to do some investing. They will have some knowledge behind them in knowing how the process works. Otherwise, your child may want to become involved with purchasing stocks and be deterred because they have no idea what it is all about. There is a load of information on the Internet about stocks, mutual funds and trading that your child can have access to. Make sure you sit down with your child during this research process so you are available to answer any questions that may arise. With having some knowledge behind them, it will give them a better foundation to start on should the time ever transpire.

If the idea does not reach out and grab your child's mind. Try again at a later date. This is a lot of information for a youngster to grasp. Little talks here and there can be more effective than one or two major talks in which your child can become uninterested. Everything can be handled in small doses. Make savings a fun project not a chore and your child will eventually see the goodness that you instilled in trying to teach them good habits.

# Chapter 15

## Let's play dress up

No offense to anyone reading this book, but this chapter is for many of my female readers and men who fit into this category. Firstly, many men do not care as much about the brand of clothing that they wear nor are as status quo as many women. Many men do fit into the status quo category and would never be caught dead in a thrift store purchasing used clothing that someone else has worn. Women please take note that the information revealed in this chapter could be very helpful to you in changing your shopping habits regarding what type and style clothing you purchase and also the quantity thereof.

Going to the mall has always seemed like a fun thing to do, especially when we have an opportunity to spend what little money we have in cash or available credit. When you need to cut back because of your "desire" and "motivation" to free up some cash flow, going to the mall to shop is not always a good idea. Many of your mall stores have a higher rent than some of the strip malls do and they need to increase the cost of their products to increase their revenue to be able to meet their own financial obligations. Even if a major anchor store (Macy's, JCPenny, Sears, etc.) has a sale advertised in the local newspaper, their sale may not be in the price range you would like it be. Unfortunately, while growing up, we learned that going to the mall was a fun thing to do. And most of the time it meant spending money we didn't really have or need to spend.

During selected holidays throughout the year a store may advertise a President's Day sale, holiday clearance sale and so on. These are the type of sales that makes it worth shopping for something new. If you watch the sale prices (seventy percent off is a great place to start), you can pick up some great deals. I usually go to a major retailer for their President's Day sale to pick up some great deals. This is usually the only time I shop there during the year as I feel I cannot afford their prices otherwise. I shopped for myself and picked up ten items of clothing for less than forty dollars. I was amazed at what I had saved, but still felt a little guilty for dipping into my rainy day/pay extra on the bills money.

Remember that just because there is a sale that does not give you an open invitation to go out and spend money. Be careful of what you spend and how you spend it. Ask yourself the question of whether you really need this and if not do something different. Do not tell yourself that you are just going to go look and not buy anything because the inevitable will happen. You will find something that is irresistible and a must have. Then you begin thinking that you really did need what you bought and that you will be more careful next time. Bologna! Look in your closet and your bureau drawers and pick out everything that you have not worn in eighteen months. Okay. So we bought all of these new clothes to replace all of the clothes that we haven't worn in ages. You need to slow down and look at the big picture. Having that many clothes is unnecessary and selfish.

If your clothes are slightly worn and do not fit you any longer, go through them and take them to a consignment store to try and get something for them. If they won't accept them, put them into the rummage sale you were planning on having and worst-case scenario you could donate them to the local Salvation Army or similar store. Just remember to get your donation slip that you can use on your taxes for a write-off next year.

The Salvation Army or another thrift store is also a good place to pick up some great deals. Keep an open mind on this subject because this is something you may have never done before or have refused to do. When you were growing up, how many times did you borrow your sister's sweater or mom's shoes? This is no different…you just have a much wider selection to choose from than when you "borrowed" (they suddenly became yours because you never returned them) them when you were younger.

Recycling is an important factor in our lives. If you are earth friendly, utilizing this concept may be just for you. This does not mean you always have to buy used, but why not check it out and see if you might just find something that you like. It could be something you fall in love with or exactly what you have been looking for that is no longer available in the department stores. Used clothing may fit a little bit better because it has been worn in and may give you a comfortable feeling. By keeping an open mind, you have more options with your budget and have the opportunity to purchase more items than if they were purchased brand new.

# Chapter 16

### Taxes...Do you know what you really should know?

**I** believe that most Americans will agree that tax time is their least favorite time of the year. Unfortunately, it is something that cannot be avoided and if you do so, you will be penalized for it if you owe taxes to the IRS (Internal Revenue Service). This may be just a minor inconvenience to those who only need to file short form or 1040EZ. The filers who need to file long form often find that there aren't many things worse than the April fifteenth deadline. I know this personally from my own experiences and have put off getting my own taxes done until as close as a week before the April 15th deadline.

By having a personal CPA (Certified Public Accountant), this required task could be less compelling than trying to accomplish this on your own. Most CPA's are more knowledgeable than the average filer as they must keep up on all of the ever-changing state and federal tax laws. They are usually aware of additional or different deductions that you can take that you were unaware of and something you may want to deduct is not permissible by the IRS (Internal Revenue Service). I have never been a fan of the tax prep companies that herd you in and out like cattle so they can file as many returns they can each day. They are not company trained to spend extra time on your tax preparation to maximize your refund. You are basically allotted a slot of time for the preparer to gain what information is needed. Some of their agents are not as knowledgeable as you would personally like. I have a "tax man" who still makes house calls and willingly allows whatever time is necessary to gain and retain any information that is pertinent to my tax return. Since switching from the cattle driven tax prep company, I have legally maximized my returns from where I used to have to pay the IRS some years to receiving a quality refund. I have comfort knowing that my tax return is in good hands and do not worry that something has been overlooked and that it is completed correctly.

If you file your own return, be aware of the tax laws and all of the items you can deduct or claim during preparation. When having your taxes done by a professional, it is always important to ask question on whether or not something is claimable. No question is a bad question because you cannot secure a satisfactory return if you do not have the knowledge of whether you can claim something or not.

During the year, keep important receipts. They can range from doctor visits to charitable donations to work related expenses. Find a box or a file to put these receipts in during the year so you will have them when it gets close to calling your CPA or preparing your own. If the IRS audits you (which they may) regardless of whether your return is correct or not, you will need to produce all documentation to prove that the deduction you took was legal.

Right before and during tax time each year, the Internet is a great place to find out about current allowable deductions and any tax laws that may have changed since the year prior to your current filing. If you are unsure about a certain deduction, the best thing to do is to ask a professional. If you file your return with a mistake in the government's favor, you will owe that amount back to the IRS plus any interest fees for not having completed your return correctly.

Below are a few examples of items that can be deducted on your return, which may be forgotten.

1. Out-of-pocket Charitable Contributions

It is hard to overlook the big charitable gifts you made during the year by check or payroll deduction. But the little things add up, too, and you can write off out-of-pocket costs you incur while doing good deeds. Ingredients for casseroles are regularly prepared for a nonprofit organization's soup kitchen, for example, or the cost of stamps you buy for your school's fundraiser count as a charitable contribution.

2. Student Loan Interest Paid by Parents

Until recently, if parents paid back a student loan incurred by their children, no one got a tax break. To get a deduction, the law held that you had to be both liable for the debt and actually pay it yourself. But now there is an exception. If your parents pay back the loan, the IRS treats it as though they gave the money to their child, who then paid the debt. So a child who is not claimed as a dependent can qualify to deduct up to $2500 of student loan interest paid by their parents.

3. Moving Expenses to Take First Job

Here is an interesting contradiction. Job-hunting expenses incurred while looking for your first job are not deductible, but moving expenses to get to that first job are. And you get this write off even if you do not itemize. If you moved more than fifty miles, you can deduct the cost of getting yourself and your household goods to the new area, including mileage (check with the IRS for the current calculation), plus parking fees and tolls for driving you own vehicle.

4. Child Care Credit

A credit is so much better than a deduction-it reduces your tax bill dollar for dollar. So missing one is even more painful than missing a deduction that simply reduces the amount of income that is subject to tax. It is easy to overlook the childcare credit if you pay your child care bills through a reimbursement account at work. Until a few years ago, the childcare credit applied to no more than $4800 of qualifying expenses. The law allows you to run up to $5000 of such expenses through a tax-favored reimbursement account at work. However, up to $6000 can qualify for the credit, but the old $5000 limit still applies to reimbursement accounts. So if you run the maximum $5000 through a plan at work but spend more for work-related childcare, you can claim the credit up to an extra $1000. That would cut your tax bill by at least $200.

5. Estate Tax on Income in Respect to a Decedent

That sounds complicated, but it can save you a lot of money if you inherited an IRA from someone whose estate was big enough to be subject to the Federal estate tax. Basically, you get an income tax deduction for the amount of estate tax paid on the IRA balance. Let's say you inherited a $100,000 IRA and the fact that the $100,000 was included in your benefactor's estate added $45,000 to the estate tax bill. As you withdraw the money from the IRA and pay tax on it, you also get to deduct a proportional amount of the estate tax paid. If you withdraw $50,000 in one year, for example, you get to claim a $22,500 itemized deduction on Schedule A.

6. State Sales Tax

This write-off is primarily for those who live in states that do not impose an income tax. You must choose before deducting state and local income taxes, or state and local taxes. For most citizens of income-tax states, the income tax deduction usually is a better deal. IRS has tables for residents of states with sales taxes showing how much they can deduct. These tables are not the last word. If you purchase a boat, vehicle, or airplane, you get to add the state sales tax you paid to the amount shown in IRS tables for your state, to the extent the sales tax rate you paid doesn't exceed the state's general sales tax. The same goes for home building materials you purchased. These items are easily overlooked. The IRS even has a calculator on its web site to help you figure out the deduction, which varies by your state and income level.

7. Reinvested Dividends

This isn't really a deduction, but it is a subtraction that can save you a lot of money. This is the break former IRS Commissioner Fred Goldberg told Kiplinger's that lots of taxpayers miss. If, like most investors, you have mutual fund dividends automatically invested in extra shares, remember that each reinvestment increases your "tax basis" in the fund. That in turn, reduces the taxable capital gain (or increases the tax-savings loss) when you redeem your shares. Forgetting to include the reinvested dividends in your basis-which you subtract from the proceeds of sale to pinpoint your gain-means overpaying your tax.

8. Refinancing Points

When you buy a house, you get to deduct points paid to obtain your mortgage in one lump sum. When you refinance a mortgage, however, you have to deduct the points over the life of the loan. This means you can deduct 1/30$^{th}$ of the points a year if it is a thirty year mortgage-that is $33 for each $1000 of points that you paid. It may not seem like much, but it adds up. Also, in the year you pay off the loan (because you sell the house or refinance again) you get to deduct all of the points not yet deducted, unless you refinance with the same lender. In that case, you add the points paid on the latest deal to the leftovers from previous refinancing and deduct the expense, which is pro-rated over the life of the new loan.

9. Jury Pay Paid to Employer

Some employers continue to pay employees' full salary while they are doing their civic duty, but ask that they turn over their jury fees to the company treasury. The only problem is that the IRS demands that you report those fees as taxable income. You have always had a right to deduct the amount so you were not taxed on money that simply passed through your hands.

10. College Tuition

You won't find this one on the forms, either, but you may qualify to deduct up to $4,000 you paid in college tuition in 2006 for yourself, your spouse or a dependent. This break can pay off if your income is too high to qualify to claim the Hope or Lifetime Learning credit. For 2006 returns, the deduction is taken on line 35 of the Form 1040, the line for the domestic production deduction. Write "T" to the left of that line. If you're claiming the production break, too, write "B" on the dotted line and attach a breakdown of how much you're claiming for each. You also get to claim this deduction regardless of whether you itemize.

11. Health Insurance Premiums

Health insurance premiums, along with long term care premiums, get added to your unreimbursed medical expenses and will be deducted to the extent that they exceed 7.5% of your AGI (Adjusted Gross Income). But self-employed individuals can deduct 100% of their health insurance premiums if they're not covered by an employer-paid plan. You don't even have to itemize to take this one!

Audits are a concern of many Americans who file income tax returns. If you complete your return properly and honestly, you should not have any issues. Below is some information to consider in helping to prevent an audit.

1. Cross your T's and dot your I's

The easiest way to avoid an audit is to ensure your return is accurate. The best way to do that - have it prepared by a certified tax professional who knows your situation. Now this may seem like common sense - a few mathematical errors alone won't get you into trouble - but several mistakes can imply that your return is a sloppy one and in need of a second look.

2. How you generate income does matter

There are also certain professions that attract more attention. For example, those who are self-employed or receive large amounts of cash (like servers and taxi drivers, for instance) are always at risk for an audit as opposed to those who receive a steady salary. Since the IRS already believes that most under-reporting occurs among the self-employed, they are on the lookout for red flags, such as taking home office deductions. It's important to weigh the risks. You can talk to your tax professional about the best option for you.

3. Alimony is not free and clear

Not reporting taxable alimony payments received can cause big trouble. The IRS now matches deductions for alimony from one spouse with the alimony income reported by the other.

4. Be accountable for automobile logs

One of the most commonly audited items are automobile logs for self-employed persons or employed individuals using their cars for business purposes. The key to protecting yourself here is to keep detailed records of your mileage on a daily basis. Try to keep a log that tracks the beginning and ending odometer readings, location and reason for the trip. Your tax professional can then help you determine how much of those miles qualify for business use.

5. Limit itemized deductions

Large amounts of itemized deductions can make you a likely target for an audit. Even though you should claim every deduction the law entitles you to, making tax deductions that exceed the averages for your income can put you at risk. For example, many people make large charitable donations in relation to their incomes. If you fall into this category, keep track of the date of the donation, the receiving organization and the amount of the contribution. Try to pay with a check whenever possible so you have a record, and make sure you obtain a receipt from the charity for donations exceeding $250. If audited, you should be prepared to also offer the IRS the original value of the items donated (non-monetary) and their fair market value. In general, the larger the donation, the more paperwork the IRS requires.

6. Take care when selling stocks or bonds

The IRS receives a 1099 form noting the sale price of your stocks or bonds. So it's best to claim capital gains in full and back out your purchase price on another section of your return. For example, if you buy stock for $2,000 and sell it for $5,000, you may think it makes sense to claim only $3,000 in capital gains. However, this will throw up a red flag. Instead, claim the full $5,000 and back out the $2,000 purchase price somewhere else on your return. A tax professional can help you complete your return accurately to avoid any mistakes.

# CHAPTER 17

### Scamming your life away...Watch out for the bad guy!

The Internet can be a great source for information, but it can lead you to have many troubles if you are unaware of the possible threats that lurk right in front of you just waiting for you to make the right move for it to sneak right in and get you. The scams are designed to catch your attention and catch you off guard. Remember that if it looks too good to be true, it probably is. Be cautious of how you deal with your personal information. Your email address, home phone and cell phone are as confidential as your social security number and banking information. Never give out any information to anyone or any company unless it is materially necessary to transact a legal business deal.

I personally have seen scams that have been sent through email and personal phone calls. I have seen scams that I have read regarding an advertisement that I had listed on craigslist (craigslist.org) for an item for sale. In the email the Sender is interested in purchasing my item, and they have agreed to the terms of my listing. They clearly state that if I furnish my address, they will send me a check to cover the cost of the item, shipping fees and inconvenience fees for selling the item. They go on to say that a Cashier's check will be issued for a bit more than what is owed and would I please refund the difference in a Cashier's check issued to them. I am then asked to send the item immediately because the check will also be sent immediately.

This is a bad business practice and it should immediately raise your eyebrow to the fact that something is wrong. A reputable person or company would never ask you to do this until you have received their payment and the check has cleared the bank. Unfortunately, in this situation you could lose your item and it could cost you the amount of money you sent to them for their need to be paid the difference in the check before your transaction is complete. If this happens to you, you should contact the local authorities immediately even though you may never have the opportunity in getting your money or item for sale back.

I have deleted many emails claiming that I have won the English Lottery and the African Lottery. How can you win something like this if you have never submitted an entry? Criminals like this request your personal banking information such as routing numbers and account numbers so they have access to your account. When you least expect it, your bank account will reflect a withdrawal instead of the million-dollar deposit that you were waiting for. Many times, the bank is unable to trace this withdrawal once it leaves the banking system due to the withdrawal bouncing from bank to bank, which most likely is out of the country. Once the money leaves the country it usually is never recovered. The bank may not reverse the money you lost because you intentionally gave out your personal information when you should not have done so.

Phone calls to land lines and cell phones are not protected from scams either. Always keep in mind that your financial institution, whether it is your personal bank or credit card company, will never call you or email you and ask you for your account number, social security number or a password. They already have all of that information on file if it is needed. Remember back to the last time you spoke with them on the phone in which you generated the call. For security reasons and to protect your account, they always ask to whom they are speaking to and what your password is. It is their process to be certain that you are identified as the owner or co-owner of the account.

By giving out your protected information, you could cause major detriment to your life and possibly open up the opportunity for Identity Theft to occur to you.

Surfing the Internet can be a lot of fun but it can turn severely against you if you are not careful. I have learned to be very cautious about filling out questionnaires, surveys and "gotcha" websites.

So, remember that this information is extremely valuable and protect it. Without guarding it, as I have stated before, it could be very detrimental to you.

# CHAPTER 18

**Is your credit important to you?**

There may be a time in your life when you need credit. Purchasing auto insurance is a necessity if you own a vehicle. The rate you are quoted and will need to pay is based on your credit report/score. If your credit report shows that you are a credit risk, what company wants to take the risk that you may not pay your premiums or be responsible with your vehicle? That is an easy question to answer. The auto insurance carrier wants to get paid for their service the same as you want to get paid from your employer. The higher your credit score (your age and zip code has a factor in your rates as well) the better the rate the insurance company can offer you.

If you do not have good to excellent credit, you may not have an opportunity to buy a house on your own. You may need to have a friend or family member co-sign for you. This situation may be difficult due to the risk that the co-signer is undertaking. Buying a house is not a fifty dollar deal that can be paid in an extremely short period of time and your co-signer is responsible for the note on the house until it is paid in full by making all of the payments, refinancing or selling the house. Since the recession that hit the US recently, most banks are not taking the risks that they took several years ago. Banks are requiring anywhere from 3 ½% to 5% or more as a down payment on a house. Financing terms have also changed where stated loans have pretty much disappeared and 80/20 loans are non-existent. Many banks got %stung%with what the media called the %sub-prime market%and %predatory lending%(The lending was not predatory; it was the people who made the loans that were the predators.) and due to the fallout of so many foreclosures, terms had to be changed and qualifications became much more difficult to meet.

# Chapter 19

## Turning the nest into a castle

Remodeling projects can be quite expensive. There are so many remodeling companies out there that are willing to update your home at a substantial cost to you. Many handymen can do the same thing and charge you a fraction of what many of those companies might charge. When hiring someone to do a major project for you, inquire as to if they are licensed and bonded in the event that something goes wrong. If something was promised a certain way and not delivered to you, express your concerns and insist that the project be fixed to your satisfaction or ask for a discount if you are able to live with the error. If you are paying for it, you should be receiving exactly what you paid for. Why should you have to settle for something less when you are expected to pay for the work? Get what you pay for! Remember to keep your budget in mind when planning the projects that you want to do. Everything does not need to be done at once. Schedule your projects around your budget so that it does not drain you financially.

In the event that you are creative and handy enough to do the work yourself, keep your budget in mind. Many of your major home centers such as Menards, Lowes and Home Depot run sales on a regular basis. If you research the ads for these stores, you will get an idea of how they do their advertising and many of their sale items run in segments. The segments may run every two to three weeks where that certain item you are looking for goes on sale again. The clearance area of these stores can be very useful in doing some of your projects where you can save a decent amount of money. Several years ago, my husband and I were interested in putting new siding on our house. We looked at the prices of siding at each of the home centers and decided that it was a little too pricey to do our project at that time. About two weeks later, Lowes discontinued one of their brands of siding and we were able to redo the house for less than one thousand dollars including all of the materials we needed. The following year, the same thing happened and we were able to redo one of our investment properties located out of state for about the same amount of money. I wanted to put hardwood floors in my kitchen but was adamant that I wanted a cherry wood, which I knew would cost me more than I was willing to spend. After waiting for a few months because the cost was too great, I got lucky because the flooring went on clearance and I saved myself over two hundred dollars.

When shopping at the major home centers, keep your eyes open for store coupons, because each of these home centers will accept the other's store coupon at their store. The home centers usually will price match for the same item and give a percentage off the difference for shopping at their location. When I am in need of carpeting, I watch the ads at Menards for any carpet to go on sale; I then take the ad to Lowes where they will price match plus give an additional ten percent off. I saved over seventy dollars on a piece of carpet by doing this. I was amazed at how well I had done and was also able to stay within my budget while making a good choice.

There are other items and many departments within these home centers that have great deals whether it is a good sales price or a clearance item. You just need to be certain that this is a project that you are willing to take on, it is something you want to do right away, and that you can fit it into your budget without ruining you financially.

By doing many projects on your own such as painting, carpeting, tiling, landscaping or even something minor, can save you a lot of money. You will not be paying someone else to do the labor and will have additional money for possible future projects.

# CHAPTER 20

### So, you want to tie the knot!

You have found that special someone or one of your children has and have decided to make the commitment of marriage. You know that weddings can be expensive and if you have to pay for it yourself, it can be overwhelming. Take the stress out of it by doing some super planning. There are ways to still have a nice wedding and keep the budget at a reasonable price.

Church weddings can get to be expensive by the time you make your donations and rent a hall for the reception afterwards. Sometimes, finding a minister who will come to your location (such as a park or a family home) can be more inexpensive than to have the service in a place of worship. Using a park or a family member's backyard can save money on renting a hall.

If this is something you want to do, recruit some family members to help with the preparations for the reception. Instead of hiring a caterer, prepare your own buffet style service food. Just by making your own potato salad for a gathering of fifty to seventy-five people could easily save you fifty to sixty dollars. You will need to refer to the shopping chapter to show you how to save money by purchasing groceries if you are uncertain as how to.

It is your reception, so you can choose how and what you want to serve. Sometimes, cold buffets go over very nicely depending on who is invited and what your budget calls for. Cutting and preparing fruits and veggies yourself can also save quite a bit of money. And if you have willing hands to help out, take advantage of the free labor to stay within your guidelines. Paper products can also be more inexpensive at dollar stores, party stores and warehouse outlets. Most of these types of stores offer decorations for sale as well. Be creative and look on the Internet to see if you can get some different inexpensive ideas to help make your gathering a success. Making your own ribbon decorations or ribbon flowers not only can be a lot of fun, but also may be less expensive at the same time. Making your own wedding favors for the guests can save a significant amount of money as well. You will just need to manage your time to be able to do the items needed for preparation.

When looking for a tuxedo to rent, shop around until you find the best package deal. Some formal wear shops will offer a free tuxedo to the groom if a certain number are rented for the attendants. If it is not a fancy ceremony, a nice suit can be substituted for the tuxedo. That you may already own. Bridal gowns can be much more complicated, especially when the bride-to-be is envisioning that fairy-tale dress she has always dreamed about. This can be possible if you are willing to do some clearance shopping so that you can stay within your means. Many bridal shops will clearance out certain styles after having them for a certain amount of time. Sometimes borrowing a gown from a family member will satisfy the bride, particularly where a budget is a concern. We always hope that this is a dress we will only need to wear once in our lifetime and it is always nice to hand down from one generation to another.

You do not need to go into debt just to get married. An inexpensive reception will give you more money towards your honeymoon. The trip to Las Vegas or a cruise to the Caribbean may be more feasible if you cut expenses elsewhere. Remember that starting out a new life together should be smooth and not overwhelming with the expenses incurred from getting married.

# CHAPTER 21

## Miscellaneous things to keep in mind

### Recycling

There are many different ways to find money that is right in front of you. It may not be cash yet, but could be turned into cash without much effort. Aluminum cans can be saved in large plastic garbage bags and taken to the recycler and turned into money that could go into savings. It can also be used to send an additional payment into a monthly creditor that you owe. Many states recycle plastic water and soda pop bottles for as much as five cents to ten cents per bottle. The side of the bottleneck will show what the redemption value is and if you save enough of them, it could be a small kitty towards a necessity.

### Uncle Sam can help

Depending on your family size and monthly income, your family could be eligible for local, state or federal aid. If you do not need it, do not access it. But if you do need the assistance because you cannot get by without, check into it to see if you qualify. There are many programs available from reduced/free school lunches, food stamps to first time homebuyer's assistance. The programs were designed to help the people who need short term help but not to live on indefinitely. Check with your local municipality to see if you may qualify.

### College money

Sending your child to college can be a draining experience both financially and emotionally. It is best to start early (usually no later than the eighth grade) to look for and apply for scholarships to help defray the cost of tuition. (fastweb.com) and (straightforwardscholarships.com) are good sites to get the ball started on scholarship searches.

When searching for student loans, make sure you do your homework and find the best rates available. The difference between a percent or a percent and a half could mean paying hundreds if not thousands of dollars more in interest than what is necessary. Like anything else you purchase, shop around for the best product before you sign on it.

**But is it used or new?**

When it is time to buy something that you really need, you should look at the big picture in deciding whether to buy it brand new or if a used item will suffice. You may be on a tight budget, but find that you need to purchase a new couch because the one you have now is out of life and beyond use. There is nothing wrong with purchasing a used couch to tide you over until you have the money in your budget to purchase something brand new. You can locate a decent couch by checking with family and friends or you can do some in depth research in locating something that will appeal to you a bit more. The local newspapers usually run ads for used items for sale and there are some local online sites such as (craigslist.org) that offer items as well.

Purchasing something a bit more personal may need to be purchased new instead of used. These items may include curling irons to personal undergarments to shoes. Some people have idiosyncrasies and would never consider using something that someone else has had or worn where other people find no problem in it. It depends on how well you want to make your budget work and how much "desire" and "motivation" you have to keep adhering to it. Most all used garments are washable and are sometimes more comfortable than buying brand new.

**Time to clean out the garage**

Your garage may not be the only thing you may need to clean out. I am sure the closets have some extras in it that are no longer needed as well. So, you are thinking about just throwing it all away because it seems easier to do. By having a yard sale (garage sale), you can recycle your "garbage" into someone else's treasure and make a little bit of money doing it. If you only have a few worthwhile items to sell, try using a local newspaper or a free service through (craigslist.org) to get rid of it and add to your savings. (Ebay.com) is another good website to utilize if the item is sellable, shippable, and in good enough condition to do so. Why not make money on your junk and put it back into your budget to help supplement the cost of other things you may need.

**Coupons can help too!**

As a grocery shopper you may not have many options on where to purchase your groceries. This is where coupons can come in handy for saving your budget. Most local newspapers supplement the Sunday paper with local advertising and also coupon booklets. By clipping the "cents off" coupons that you will use, it can save you a few dollars to several dollars on your shopping trip.

Grocery coupons are not the only coupons that you can utilize to help save money on purchases. There are coupons out there for virtually everything from oil changes to gutter replacement to restaurants. Coupons can be found in newspapers, magazines to finding a a mailer in your mailbox. At one point in my life, I would not use a coupon to save my soul; I was afraid that I would turn into my mother. Guess what? I grew up very similar to her! But now, I will clip the coupons that I think I might use and go through them periodically to eliminate the ones that I decided not to use and to remove anything that may have expired.

**Shopping for electronics**

Purchasing anything electronic can be spendy and a lot of thought should be put into the decision. Shop around, as there are so many different places to purchase these items. During the holidays, many stores have incredible sales on these types of items and if you watch the local ads, you may find something with a great price on it. At one time, trying to find a laptop under four hundred dollars was near impossible unless you signed up for a year contract with an Internet provider. Do you remember that? Recently, I saw a decently equipped laptop at a local retailer (microcenter.com) for just below three hundred dollars. Many of the stores are very competitive on pricing and can be quite confusing on the bottom line. Researching on the Internet is a good tool as you can do comparison shopping from Website to Website. Don't forget to keep Radio Shack in mind for some of your shopping comparables. I have purchased several GPS (Global Positioning Satellite) from this chain as I found that they had the best deal for less than one hundred dollars. Keep in mind though that if you don't need it, (want is not the same thing) don't buy it. Also, the cheapest is not always the best deal. You get what you pay for and electronics is an area where it is hard to skimp on certain things.

**Gasoline keeps us going**

The cost of gasoline has changed so much over the years, but has hit us hard over the last few. We always seem to complain about it when we have to fill up our tank and then put it into the back of our minds for the next time we might share the information with others. It seems to be a popular topic amongst many.

I usually fill my tank at convenience stores that accept "cents off" coupons for gasoline and will accept competitor's coupons as well. I have heard many people tell me that they will drive several miles out of their way to save two or three cents per gallon of gas. This is crazy. The money just saved from the gasoline purchase was spent on the drive to and from the location that was out of the way. So that is no money saved. I will stop for gasoline at a station that is cheaper if I am already going to be going by it, but I will not go out of my way to do so.

While on this subject of gasoline, there is a point that should be made regarding our children and their friends. Your child works hard to keep gasoline in their tank so they can drive all over the place. The next thing you know they are picking up their friends and driving them everywhere as well. Your child (kind-hearted as they may be) gladly does this because their friends do not have a ride so they can get together. All of a sudden, your child is begging for money to put gasoline in their tank because they are on empty. This is where we need to stop. Inform your child that they need to take responsibility for lugging their friends around and need to start collecting from each and every friend who needs a ride. Their friends will eventually cough up some cash when your child stops picking them up and dragging them all over.

LaVergne, TN USA
26 May 2010
183949LV00005B/3/P

9 781452 013374